CIRIA C664

# Iron and steel bridges:

## condition appraisal and remedial treatment

*Prepared under contract to CIRIA by Gifford WSP*

| | |
|---|---|
| **G P Tilly** | Gifford |
| **S J Matthews** | WSP |
| **D Deacon** | Steel Protection Ltd |
| **J De Voy** | Gifford |
| **P A Jackson** | Gifford |

CIRIA  *sharing knowledge* ∎ *building best practice*

Classic House, 174–180 Old Street, London EC1V 9BP
TELEPHONE 020 7549 3300     FAX 020 7253 0523
EMAIL       enquiries@ciria.org
WEBSITE   www.ciria.org

**Iron and steel bridges: condition appraisal and remedial treatment**

Tilly, G P, Matthews, S J, Deacon, D, De Voy, J, Jackson, P A

*CIRIA*

C664        © CIRIA 2008 (revised)        RP721        ISBN: 978-0-86017-664-0

**British Library Cataloguing in Publication Data**

A catalogue record for this book is available from the British Library.

| Keywords | | |
| --- | --- | --- |
| Transport and infrastructure, facilities management, health and safety, knowledge management, materials, materials technology, regulation, site management, sustainable construction, whole-life costing | | |
| **Reader interest** | **Classification** | |
| Asset management, maintenance and upgrading of iron and steel bridges | AVAILABILITY | Unrestricted |
| | CONTENT | Advice/guidance |
| | STATUS | Commissioned, committee-guided |
| | USER | Asset managers, designers, contractors, owners |

Published by CIRIA, Classic House, 174–180 Old Street, London, EC1V 9BP

# Summary

This book aims to meet the requirements of those having a general knowledge of bridge engineering and asset management but requiring information about the performances and specific requirements of iron and steel bridges.

In the UK there are some 22 000 iron and steel bridges. They have a very important role as part of the infrastructure and need to be maintained to meet modern requirements of strength and serviceability which, in most cases, far exceed what was considered in their original design. Many are heritage listed as they have great significance in industrial history, for example Ironbridge in Shropshire, and their maintenance is subject to requirements of legislation.

The historic development of materials, from cast iron to high strength steel, is described as it is important to have this knowledge when managing and assessing bridges of varying ages. For the same reason, the development of structural form is described, from the early arches having connections developed from carpentry, to modern stiffened plated construction having welded joints.

Guidance is provided on the relevant statutory requirements that must be followed in the care and maintenance of bridges as these have a profound effect on design and planning of projects.

Asset management is explained, including bridge management systems, preventative maintenance, and best practice.

The defects and processes of deterioration that typically occur in iron and steel bridges are described and illustrated by photographs.

There are substantial sections on inspection, testing and monitoring. These explain the strategies of inspection, the range of available laboratory and field tests, and methods of monitoring the behaviour of bridges as an aid to diagnosis or surveillance preceding remedial action or replacement.

Methods of structural assessment, from low-level appraisal to non-linear analysis, are given. Topics considered include: codes of practice, treatment of damage and deterioration, management of under-strength bridges, and financial liability.

The design and execution of remedial works are considered in three sections; general principles, repair, and strengthening. Repair is considered as the restoration of a structure to its design strength (or serviceability), strengthening is concerned with raising the load carrying capacity to above the original design value. Strengthening projects have been mainly concerned with the older bridges but also include more recent bridges such as the Docklands Light Railway. The examples described are of rail, canal and highway bridges.

Protection of iron and steel from corrosion is crucial to long life and low maintenance. The historic development of paint systems and current health and safety issues are described. There are sections on modern systems of protection, including inspection and repair, and treatment of ironwork, buried structures and complex configurations.

# Acknowledgements

The authors acknowledge information kindly provided by Ed Proctor (Nuttall), Derek Hughes (Taylor Woodrow), Simon Clubley (Mott MacDonald), Martin Lorrimer (Dundee City Council), Ian Chapman (TfL), Pat Warrener (Hampshire County Council), Graham Slade (WSP), Tony Oakhill (Gifford), Will Duckett (Gifford), Barry Mawson (Rtd) and John Holly (EH Services Ltd).

A significant contribution to Chapter 6 was made by Dr Martin Ogle (TWI).

## Project funders

The need for this work was identified by the Bridge Owners Forum. CIRIA and the Bridge Owners Forum are grateful to the following for providing the funding for this project:

Bridges Board, Department for Transport

Network Rail

CIRIA Core Programme Sponsors

Transport for London

Health and Safety Executive

British Waterways

## Research contractor

Gifford WSP

## Authors

**Graham Tilly BSc(Eng), DIC, ACGI, PhD, CEng, FICE**

Graham is an emeritus director with Gifford. He has some 35 years experience in bridge engineering and research with expertise in materials, fatigue, whole life costing, and conservation. He has written some 80 papers and several books and was the lead writer of the guide.

**Steve Matthews MSc DIC C.Eng FIStructE MICE MIoD**

Steve is senior technical director for WSP UK, bridges and civils. He has 30 years experience at home and overseas for public and private sector clients, working for large multidisciplinary and small specialist consulting engineers, steelwork fabricators, and research organisations. He specialises in structural steelwork, composite construction, design and construct and bridge management.

**Paul Jackson BSc PhD CEng FICE FIStructE**

Paul is a Gifford technical director with 30 years experience of bridge design, analysis, construction and research and he serves on bridge code committees. He has particular expertise in assessing the real strength of structures. His main contribution was in the assessment of structures.

**Julian De Voy, BSc PhD CEng MICE**

Julian De Voy, formerly a Gifford associate, has more than 15 years' practical experience of planning and undertaking inspection, testing and strengthening contracts on metallic structures. He contributed the section on inspection. Julian is now a technical manager with Grontmij, Solihull.

**David H Deacon CSci, FICorr (Hon), FTSC**

David is a director of the Steel Protection Consultancy and has had 40 years experience as a paint technologist. He has been employed on the Forth Road and Rail Bridges, Tower Bridge, Blackfriars, Southwark, Hammersmith and Lambeth Bridges in Central London, as well as bridges in Asia and the Middle East. He contributed the section on coatings.

### Steering group chair

| | |
|---|---|
| Mr J Clarke (chair) | BRB (Residuary) Limited |

### Steering group

| | |
|---|---|
| Mr D Ashurst | Arup |
| Mr B Bell | Network Rail |
| Mr G Bessant | TfL |
| Mr P J Brown | Oxfordshire County Council |
| Mr E Bunting | Department for Transport |
| Mr M Bussell | Arup |
| Mr M Collings | Gifford |
| Mr K Duguid | Transport for London |
| Mr A Ghose | Faber Maunsell Ltd |
| Mr G Kerr | HSE (Railway Inspectorate) |
| Mr M Holford | Severn Trent Water |
| Mr R Howe | British Waterways |
| Dr. R.J. Lark | Cardiff School of Engineering |
| Mr C Mayer | Stoke City Council |
| Dr J Menzies | Independent Consultant |
| Mr A Monnickendam | Cass Hayward LLP |
| Mr T Swailes | University of Manchester |
| Mr I Wesley | Atkins |
| Mr T Wilkins | Mott MacDonald |

### CIRIA managers

| | |
|---|---|
| Miss S Connolly | Project manager to April 2007 |
| Mr K N Montague | Project director |

### Photographs

Photographs have kindly been provided by the authors' organisations or by members of the Steering group.

# Contents

\* Bridge recognised nationally as having special historic significance

## Figures

## Tables

## Boxes

# Glossary

| | |
|---|---|
| **Bridge deck** | Superstructure including running surface and excluding beams when separate. |
| **Brittle fracture** | Cleavage fracture which occurs in body centred cubic structures on (100) crystallographic planes. Brittle fractures have negligible ductility and energy absorption. |
| **Cementite** | Hard constituent of white cast iron, an iron carbide, $Fe_3C$. |
| **Characteristic strength** | Mean strength less 1.64 standard deviations. |
| **Faying surface** | Connecting faces between steel plates in a bolted joint. |
| **Ferrite** | Pure iron. |
| **Graphite** | A form of carbon. |
| **Graphitisation (graphitic corrosion)** | Corrosion of grey cast iron in which the metallic iron is converted into corrosion products leaving the graphite intact. |
| **Hoggin** | Well graded gravel having clay binder. Historically used for tracks and paths but also sub-pavement over jack arches. |
| **Half through bridge** | A bridge in which the lower chord carries the vehicular or pedestrian traffic. |
| **Huck bolts** | Proprietary bolt. |
| **Macalloy bars** | High strength steel bars commonly used for prestressing. |
| **Pearlite** | A layered structure having a pearl-like appearance under the microscope and composed of cementite and ferrite. |
| **Prepreg** | Pre impregnated fibre. Resin impregnated cloth, mat or filament in flat sheet. The resin is often partially cured to a tack-free state. |
| **Repair** | Restoration of a structure to its design strength. |
| **Scoria** | Larva-like material used for filling (bodging) defects in cast iron. |
| **Slag** | A by-product in the manufacture of iron and steel. |
| **Strengthen** | Increase the load carrying capacity of a structure to above the design value. |
| **Stringer** | Bands of slag aligned in the direction of rolling wrought iron. |
| **Through bridge** | A bridge in which the lower chord carries the vehicular or pedestrian traffic and having cross-bracing located above the traffic. |

# Abbreviations

| | |
|---|---|
| ACFM | alternating current field measurement |
| ACOP | approved code of practice |
| ACSM | alternating current stress measurement |
| BMS | bridge management system |
| CDM | construction, design and management |
| CFRP | carbon fibre reinforced polymer |
| ECA | engineering critical assessment |
| FRP | fibre reinforced polymer |
| HAZ | heat affected zone |
| HSE | Health and Safety Executive |
| HSFG | high strength friction grip |
| ICATS | Industrial Coating Applicator Training Scheme |
| ILO | International Labour Organisation |
| LVDT | linear variable differential transducer |
| MIO | micaceous iron oxide |
| MPI | magnetic particle inspection |
| NDT | non destructive testing |
| NIOSH | National Institute for Occupational Safety and Health |
| OES | optical emission spectrometry |
| PTFE | poly tetra fluoro ethylene |
| RSJ | rolled steel joist |
| SAC | special area for conservation |
| SSSI | site of special scientific interest |
| STRA | short transverse reduction in area |
| TCB | tension control bolts |
| TIG | tungsten inert gas |
| UC | universal column |
| UHM | ultra high modulus (CFRP) |
| UIT | ultrasonic impact testing |
| UP | ultrasonic peening |
| UTS | ultimate tensile strength |

## Units

| | | |
|---|---|---|
| 1 inch | = | 25.4 mm |
| 1 foot | = | 304.8 mm |
| 1 pound force | = | 4.448 N |
| 1 ton force | = | 9.964 kN |
| 1 lb/in$^2$ | = | 0.0069 N/mm$^2$ |
| 1 ksi | = | 6.9 N/mm$^2$ |
| 1 ton/in$^2$ | = | 15.44 N/mm$^2$ |
| 1 micron | = | 0.001 mm |

# 1 Introduction

## 1.1 BACKGROUND

The infrastructure of the UK including road, rail, canal, docks, and footpaths, contains some 22 000 iron and steel bridges. These bridges range in size from single-span to multi-span viaducts and are constructed using cast iron, wrought iron and steel. They date from the oldest surviving iron bridge at Coalbrookdale built in 1779. Built in 1800, Coalport Bridge nearby continues to carry vehicular traffic (see Figure 1.1). Some 12 000 of the 16 000 rail bridges were constructed before 1914 and many exceed the notional design life of 120 years given in BS 5400. Many iron and steel bridges are heritage listed and require special treatment bound by legislation. In general they have performed well, particularly in light of the increased traffic loadings and intensities, but all require regular inspections, maintenance and some require strengthening. Many have received remedial work in the past, however this is often considered dated in the light of current knowledge. Correct application of modern state-of-the-art understanding of the behaviour of iron and steel bridges, and better maintenance and repair techniques, enables these vital structures to be kept in use for the foreseeable future.

Steel bridges built in the motorway era have been designed to more modern standards but many have required strengthening or widening to meet the requirements of increased volumes and weights of traffic.

**Figure 1.1**   *Coalport Bridge*

It has been recognised that there would be benefit in drawing together into a single guidance note the combined knowledge of the principal bridge owners and engineers on the asset management of such bridges. Asset management in this context includes risk management, inspection, assessment, maintenance, repair and strengthening. Supportive tools include high level structural analysis, material testing, load testing and monitoring. Many owners already have specific procedures drawn up over many years to suit the operational requirements of the organisation. Each set of procedures is broadly similar but varies in detail and emphasis.

Previous reports by CIRIA have provided best practice guidance on the condition appraisal, assessment and remedial treatment of infrastructure for embankments CIRIA C591 (Perry *et al*, 2003a), cuttings CIRIA C592 (Perry *et al*, 2003b) and guidance for the assessment and repair of masonry arch bridges CIRIA C656 (McKibbins *et al*, 2006). These projects have demonstrated that considerable benefits will arise from synthesising the asset management procedures of different owners into a common best practice approach for all such structures.

## 1.2 AIMS OF THE BOOK

This book builds on earlier CIRIA reports and is complementary to the guidance on masonry arch bridges, CIRIA C656 (McKibbins *et al*, 2006), which has many features and requirements in common. It aims to provide infrastructure owners, their designers, contractors, asset and maintenance managers with comprehensive, authoritative and impartial best practice guidance on all aspects of the asset management of iron and steel bridges.

This book aims to meet the requirements of those having a general knowledge of bridge engineering but requiring information about the performances and specific requirements of the appraisal and remedial treatment of iron and steel bridges.

Some of the more general aspects common to all types of bridge such as asset management, risk management and the regulations associated with health and safety, the environment and heritage that have been covered in the other CIRIA publications, are briefly introduced and referenced so that readers are informed of the generalities and can obtain more detailed guidance at source.

Use of this guidance will lead to cost savings through a reduction in the deterioration of iron and steel bridges, and increased confidence of owners to apply whole life costing with improved safety and cost-effective long-term management strategies to their bridge stocks. Moreover there are great benefits of sustainability to be obtained from the maintenance of existing structures and consequential savings of natural resources that would otherwise be consumed in reconstruction.

## 1.3 SCOPE

This book is concerned with single- and multi-span iron and steel highway, rail, canal and other bridges and viaducts of all ages and having spans of 2 to 50 m. It deals with the more common structural forms:

- arches
- box girders
- plate girders
- truss girders
- lattice girders
- through girders
- suspended bridges.

And decks:

- jack arch decks
- buckle plate construction
- infill decks
- trough decks
- longitudinal timber troughs
- orthotropic decks.

Most of these components relate to wholly steel bridges while others relate to composite construction having concrete, masonry or brickwork acting structurally with iron or steel.

The requirements of footbridges, demountable bridges and movable bridges are considered within the aforementioned types. Unique bridges, for example transporter bridges and long-span bridges (spans greater than 50 m), are not considered.

This book deals with iron and steel bridges which are the most dominant metal bridges in the UK. The very few (single figures) aluminium bridges in the UK are described in Appendix A1.

## 1.4 ISSUES OF SPECIAL IMPORTANCE TO IRON AND STEEL BRIDGES

Topics of particular importance in the management of iron and steel bridges include:

- the need to understand the construction materials, their properties (including variability) and behaviour
- the need to understand the regimes of built-in stresses and their influences on overall behaviour
- the characteristics of brittle fracture and fatigue in relation to iron and steel
- the process and development of corrosion, its potential effect on structural integrity, and interactive effects in promoting brittle fracture and fatigue
- the problems posed by older designs and modes of construction that are no longer in use
- protective systems, their characteristics and useful lives. The resistance of protective systems to environmental effects
- consideration of historic and aesthetic aspects of bridges and the requirements of heritage authorities for sensitive treatment of listed and scheduled bridges and their environs
- recognition of the aims of sustainable construction through maintenance of existing structures and minimal introduction of new materials.

## 1.5 HOW TO USE THE BOOK

The structure of the book is designed to be clearly laid out so that specific information can easily be identified. It reflects the various aspects of appraisal and remedial treatment.

Chapters 2 and 3 provide information about materials, construction and the regulations that dictate what can and cannot be carried out in relation to maintenance and construction operations on the bridges. It is important to be aware that statutory requirements are constantly being updated and it is necessary to ensure that the latest information is available.

Chapters 4 to 6 deal with assessment starting with information about the types of defect to be expected, methods of detecting defects including specialist techniques enabling removal of sample material while causing minimum damage to the structure. The different levels of structural analysis, supported by supplementary load testing, can be used to enable hidden strength to be identified.

Chapter 7 looks at structural assessment including preliminary and advanced methods of numerical analysis.

Chapters 8, 9 and 10 deal with the various construction operations, maintenance (both routine and preventative), repairs to defective components and strengthening.

Chapter 11 describes the types of protective system, preparation of the iron and steel, application of the protective system and performances.

Appendix A1 deals with aluminium alloy bridges giving examples, material properties, performances and the future potential for the material.

Appendix A2 provides case studies of repair and maintenance of iron and steel bridges. These include unsuccessful interventions as well as successes.

Appendix A3 gives information about the Industrial Coatings Applicator Training Scheme (ICATS) for the training and qualification of paint operatives.

Appendix A4 contains a bibliography of useful references grouped under:

- materials and structural forms
- assessment
- examples of repair and strengthening.

# 2          Materials and structural forms

In this chapter the historic methods of manufacturing iron and steel are described, and typical material properties are given.

The common structural forms of bridges and types of deck are outlined.

## 2.1          MANUFACTURE OF IRON AND STEEL

When carrying out an appraisal and planning remedial treatment for defective iron or steel bridges, it is important to have a full knowledge of the original materials, their manufacture, characteristics, performances and the way they have been used in construction. More detailed information on these topics has been provided by Bussell (1997), Tilly (2002) and Swailes (2006).

Dates when cast iron, wrought iron and steel were in common structural usage are given in Table 2.1.

### 2.1.1          Cast iron

It became possible to produce reliable cast iron in the early 1700s and one of the first iron bridges, Ironbridge in Coalbrookdale, was completed in 1800. In the manufacture of cast iron, the molten iron produced in a blast furnace was poured into a trough called a sow with offshoots called pigs, hence the term pig iron. It was customary to mix different varieties of pig iron from different localities to achieve what was considered to be an optimum mix for a particular use. Cast iron used for structural applications was invariably grey iron made from the best quality pig iron and considered to be the most reliable.

Cast iron contains typically two to five per cent carbon. In white cast iron the carbon is present as cementite and the fracture of such an iron has a white appearance. White iron was rarely used in bridge construction. In grey cast iron most of the carbon is present as flakes of graphite, the rest is in the form of pearlite or ferrite and the fracture is grey. As cementite is very hard, white cast iron is hard and brittle. Graphite is soft so that grey cast iron is less hard, readily machineable and less brittle.

The grey cast iron made nowadays is similar to the historic cast iron but made to higher standards, and has less variable properties. Ductile cast iron is a modern material with the graphite present as spherical nodules and properties similar to carbon steel.

Unsoundness of castings was a problem that remained throughout the era of cast iron bridges and there are well chronicled cases of failures caused totally or partially by draw (blow) holes, for example Inverythan railway bridge in 1882, Berridge (1969), and the better known Tay Bridge in 1879. The earlier collapse of Dee Bridge in 1847 led to cast iron being replaced by wrought iron in design of large girder bridges. Cast iron continued to be used for short spans and arch bridges.

Cast iron can be recognised superficially by the more rounded shape of its edge details, the characteristic appearance of its sand moulded surface texture, and greater thickness of flanges, typically 40 to 50 mm. When there is doubt and the identity of the material is needed, tests can be carried out as described in Chapter 6.

A micro-section of cast iron is given in Figure 2.1. For more detailed information about the properties of cast iron see Angus (1976).

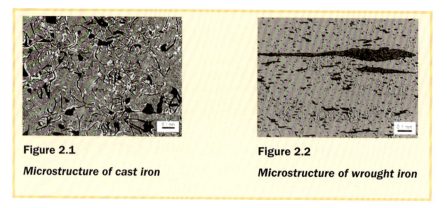

**Figure 2.1**

*Microstructure of cast iron*

**Figure 2.2**

*Microstructure of wrought iron*

### 2.1.2 Wrought iron

The puddling process where the molten iron was raked to ensure that it was uniformly exposed to oxygen enabled mixed iron and slag to be produced which could be hammered and rolled into bar. This puddle bar was cut into suitable lengths and piled into orthogonal layers to produce a 0.5 m cube of bars secured together by wire or thin bar. The cube was reheated to 1300°C, hammered and rolled into bar. This process was then repeated to produce *best bar*. Repeating the process again produced *best best bar*. Strength and ductility was improved each time the process was repeated as the stringers of slag were broken up to produce a finer dispersion with the fibres aligned in the direction of final rolling. Some banding of the matrix structure may still be retained and becomes evident in cases where the wrought iron corrodes. A micro-section of wrought iron is shown in Figure 2.2. This corrosion characteristic has a useful by-product as it enables wrought iron to be identified on occasions when there is doubt.

Wrought iron is composed of almost pure iron plus about one per cent slag. It was made in small batches and consequently had variable properties from batch to batch. This is a significant factor in modern assessment of strength. Wrought iron has a higher performance than historic cast iron, superior tensile strength and ductility. By the 1840s wrought iron was being produced in bulb channels and small I-sections. Later, it was extended to larger sections, riveted plate and lattice girders etc. The use of wrought iron for major bridges declined towards the end of the 19th century and the last major wrought iron bridge to be constructed was by Eiffel in 1884.

Wrought iron components were formed by rolling and are characterised by having constant cross-sections. They were in relatively short lengths and longer components required connections riveted with splices or cover plates.

### 2.1.3 Steel

Steel sections were available in limited sizes from 1850 but commercial supplies were very expensive. In Britain most of the native iron ores were phosphoric and production of steel from phosphoric ores became possible in 1878 when Thomas and Gilchrist invented the basic process which enabled the harmful phosphorus to be removed. The manufacture and quality of steel developed rapidly and by 1887 Dorman Long had produced a standardised range of beams and other products. The first major steel bridge in Britain, the huge Forth rail bridge, was constructed in 1890 using building techniques and workers imported from nearby shipyards.

Steel derives its mechanical properties from a combination of chemical composition, heat treatment and manufacturing process. Steel can contain up to 2.5 per cent carbon, above this value the material is classed as cast iron. Alloying of carbon to iron increases its strength but lowers its ductility, for example a steel containing 0.4 per cent carbon may be twice as strong as pure iron. The early historic steels typically contained 0.2 to 0.5 per cent carbon. In later years structural steel was developed containing 0.1 to 0.25 per cent carbon providing an altogether better combination of strength and ductility.

Steel is cast into ingots and rolled to the required sections, and it has a more homogeneous structure than wrought iron.

Structural steels produced up to the 1950s were of poorer quality than current steels and were liable to contain defects such as laminations, deformities and inclusions. Early steels were also liable to brittle fracture due to having low fracture toughness and high brittle-ductile transition temperatures. Brittle fractures can typically initiate from stress concentrations and defects. With the introduction of welding in the late 1930s this was highlighted when a bridge at Hasselt in Belgium collapsed in March 1938 14 months after construction, at a temperature of -20°C. The fracture initiated at a defect in a weld.

The extent of the brittle problems became apparent when catastrophic brittle fractures occurred in the all-welded liberty ships during the 1940s. This led to an intensive programme of research through the 1950s, and the eventual improvements in the composition and manufacture of steel and welding processes.

The first welded bridge in Britain, Billingham Branch Line, was constructed in 1934 and although no longer carrying an active railway has survived to be used by pedestrians.

**Table 2.1**    *Chronology of the use of iron and steel in bridges*

| Material | Approximate dates |
|----------|-------------------|
| Cast iron | 1780 – 1900 |
| Wrought iron | 1810 – 1880 |
| Steel | 1880 – present |

Weathering steels, introduced in the 1970s, have good resistance to corrosion and under favourable conditions do not require painting. These steels are high strength low alloy and have mechanical properties similar to grade S355 steels to BS EN 10025-1 (2004). Chapter 11 gives information about their performances under unfavourable conditions.

An example of a 15 year old bridge across rail having weathering steel beams is shown in Figure 2.3.

**Figure 2.3**     *Bridge having weathering steel beams*

## 2.2     MATERIAL PERFORMANCES

### 2.2.1     Cast iron

Historic cast iron is relatively strong in compression but is brittle and weak in tension. It is liable to have numerous types of defect mostly produced during the casting operation, see Chapter 5.

Cast iron has a non-linear relationship between stress and strain due to the presence of graphite flakes and it follows that there is no clearly defined yield point, see Figure 2.4. The presence of graphite flakes in the micro-structure of cast iron causes the embrittlement; a defect-free cast iron would still be brittle.

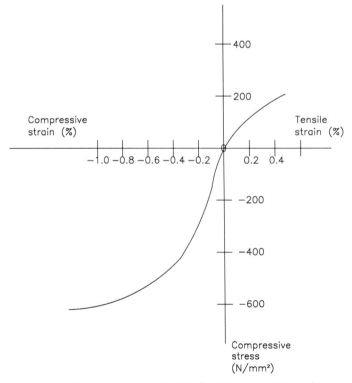

**Figure 2.4**     *Typical stress-strain relationship for historic cast iron, from Bussell (1997)*

Early cast iron I-beams having equal top and bottom flanges were used in bridges up to the 1840s. In many cases the beams were proof loaded to stress levels close to ultimate strength. Beams with bottom flanges three or four times bigger than the top flanges were widely used from the 1830s onwards when the relative strengths in tension and compression were better understood.

During the years 1830 to 1850 many combinations of cast iron and wrought iron were used in efforts to compensate for the shortcomings of cast iron in tension.

Modern cast iron containing spheroidal graphite is ductile in both tension and compression and can be used as a substitute for historic cast iron when it is necessary to raise the performance of the component in question, for example for parapets.

Cast iron has good resistance to corrosion evident in Figure 2.5.

Figure 2.5          *Cast iron bridge after some 50 year weathering without maintenance*

## 2.2.2          Wrought iron

In wrought iron, the stringers are aligned in the plane of a plate or the axis of a rod and cause the material to have anisotropic properties, the strength perpendicular to the stringers is about two thirds of the strength parallel to the stringers. Depending on the degree of working received during manufacture, corrosion of wrought iron plates can sometimes occur in planes adjacent to the stringers causing the material to delaminate.

Wrought iron has good ductility of ten per cent or more.

The research and preparatory work carried out by Robert Stephenson and his collaborators, Fairbairn and Hodgkinson, in support of the designs of the bridges crossing the River Conway and the Menai Straits not only defined the properties of wrought iron for years to come but represented one of the most significant achievements in engineering design. Their investigations included buckling, bending, deflection, friction of riveted connections, fatigue, wind forces and temperature effects. Conway Bridge was completed in 1848 and load tests were carried out to confirm the design. Britannia Bridge was completed two years later.

A particularly notable fatigue investigation was carried out by Fairbairn on riveted beams fabricated from wrought iron plate and angle-irons. The beams were 6.7 m long and 0.4 m deep. The tests were in repeated-bending involving up to three million cycles per beam and ran from March 1860 to January 1862. The object was to verify the maximum design stress of five tons per square inch (77 N/mm²) which was applied for many decades to the design of iron and subsequently mild steel structures, Fairbairn (1864).

In the 1950s about 1000 rail bridges with wrought iron beams were identified by the former British Railways as being potentially at risk of fatigue failure. A number were in the course of being reconstructed or replaced and the opportunity was taken to test the redundant beams in bending fatigue. It was found that although the beams were of widely different manufacture and age, and wrought iron had variable properties from batch to batch (as indicated in Table 2.2), they gave consistent fatigue performances, specifically their fatigue limit. In the majority of cases it was concluded that this uniformity was probably because rivet holes in the beams tended to control crack initiation and subsequently fatigue life. The upshot of this programme was that very few wrought iron bridges had to be replaced on the basis of fatigue, McLester (1988).

Wrought iron has a lower resistance to corrosion than cast iron and is very poor in a marine environment, evident in the condition of Brunel's SS Great Britain when rescued for conservation in the 1970s.

## 2.2.3      Steel

Historic steels have variable ultimate tensile strengths which may typically be in the range 28 to 32 ton per square inch (432 to 492 N/mm²). The earliest steels had an even wider range of strengths due to the difficulties in controlling the percentages of carbon and trace elements. In the absence of specific data a characteristic (yield) strength of 230 N/mm² is recommended in BD 21/01 for steels produced before 1955. When more accurate values are required it is necessary to remove samples and carry out tests to determine strength and ductility, as described in Chapter 6. An indication of the influence of carbon content on strength and ductility is shown in Figure 2.6. Typical strengths of cast iron, wrought iron and steel are given in Table 2.2.

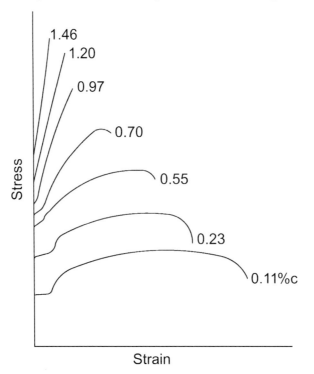

**Figure 2.6**

*An indication of the influence of carbon content in steel*

Historic steels are generally unsuitable for welding but wrought iron can be hot-welded in a forge (forge-welded). This is a difficult operation and should not to be attempted lightly.

Modern steels, having lower carbon contents, exhibit stress-strain relationships which have a clearly defined linear elastic region, upper and lower yield points, and a considerable ductile-plastic region prior to work-hardening, as shown in Figure 2.7.

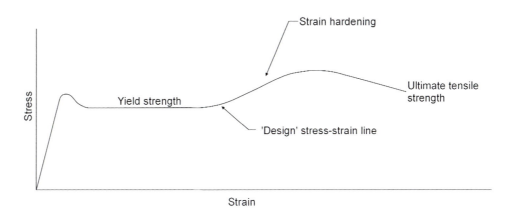

**Figure 2.7**          *Stress-strain relationship for modern steel*

Steel corrodes more readily than cast iron and requires regular painting. Exposed steel corrodes at a rate of about 0.05 mm loss of thickness per year in a rural environment and 0.05 to 0.1 mm in an industrial environment. It should be noted that heavy accumulations of rust do not necessarily indicate significant losses in sections because general rust (fully oxidised iron as opposed to local corrosion) occupies up to 10 times the volume of original material.

**Table 2.2**          *Typical ranges of UTS*

| Material | UTS range N/mm² |
|---|---|
| Structural grey cast iron | 65 – 100* Lowest<br>150 – 280* Highest |
| Wrought iron | 278 – 593*<br>309 – 386 BS 51 (1939) |
| Structural steel | 432 – 492 BS 15 (1906)<br>494 – 602 BS 968 (1962) |

\* data from Twelvetrees (1900)

## 2.3    STRUCTURAL FORMS

There are a number of alternative ways of classifying the structural forms of iron and steel bridges. One is described below.

Bridge owners such as Network Rail and the Highways Agency adopt classifications that reflect the particular features of their structures, and the way that data relating to these features is handled within their bridge management process.

### 2.3.1    Bridge types

#### Arch bridges

The early cast iron bridges were made in sections and connected together by mechanical means, for example Ironbridge in Coalbrookdale was composed of 10 half girders. The existing technology of masonry arches and timber construction was influential on the design of early bridges, and connections were carpentry style by mortise-and-tenon, dovetail, or bolted joints. An example of a bolted joint is shown in Figure 2.8. The form of cast iron arch bridges evolved to include solid spandrels, spandrels filled by circles or other decorative features and spandrels filled by lattice construction to give added strength, see Figure 2.9.

Note the poor quality and blow holes in the left hand casting

**Figure 2.8**

*Bolted joint in arch rib c1860*

Some examples of notable early iron bridges that are still in use include:

- Coalport, 32 m span, 1800, see Appendix A2.1
- Newport Pagnell, 18 m span, 1810
- Waterloo, Betws-y-Coed, 32 m span, 1815
- Mythe, 52 m, 1825.

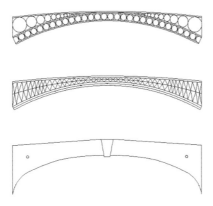

**Figure 2.9**        **Early forms of cast iron arches**

When wrought iron became available various hybrid combinations of wrought and cast iron were used in efforts to optimise construction, for example a cast iron top flange with a wrought iron bottom flange. Once painted these combinations can appear to be monolithic and can present unsuspected problems in assessment work.

One of the earliest arch bridges to be constructed in steel is at Cragside, Rothbury. It was completed in 1875 and has a span of 19.5 m. Some relatively long span arch bridges have subsequently been constructed in steel including Tyne Bridge Newcastle with a span of 162 m, in 1928, and the Runcorn-Widnes Road Bridge with a span of 330 m, in 1961.

## Truss girders

In truss girders the bending moments are resisted by top and bottom chord members. Shear forces are resisted by discrete diagonal and vertical members, as opposed to metal plates (as in plate girders) or a lattice of smaller components (as in lattice girders). Section 7.4 deals with methods of analysis.

There are many different geometric forms of truss girder, each with its own advantages and disadvantages. A particular form of truss commonly used on footbridges, the Vierendeel Girder, has only vertical members between the top and bottom chords, and carries shear forces by inducing local bending moments at the member connections. A description of the more commonly used types can be found in the *Steel designers' manual* (Owens and Knowles, 2003).

Truss girders are frequently encountered in bridge construction. They can be used for a range of spans, but have benefits on longer spans, where a combination of strength and stiffness is required to control deflection.

The typically more robust nature of the individual members of a truss girder often requires using heavier equipment to construct them.

The nodes formed where chord and diagonal members meet form traps for collection of detritus which lead to corrosion, and should be detailed in such a way that water drains away freely and protective treatment can be easily applied and repaired.

An example of a modified Warren truss is illustrated in Figure A1.3 of Appendix A1.

## Lattice girders

Lattice girders can be regarded as a form of truss having webs fabricated from a lattice or trellis arrangement of relatively small section material, usually flat bars or angle sections. Lattices have the advantage that they are lighter in weight than equivalent plated webs. They were a popular form of construction in the early 1900s. At this time, designers and fabricators were able to take advantage of the availability of smaller strip and section material, as opposed to larger wrought iron or steel plates. Construction of lattice girders is labour-intensive to produce, but quite large girders can be fabricated by relatively light equipment.

Lattices present a significant maintenance liability, because of the large number of nodes. Each intersection presents a potential corrosion trap in terms of application of protective coatings, water retention, corrosion and remedial work, as described in Section 4.2. Lattices are often found on smaller scale structures such as footbridges, and, less commonly, on larger structures, for example Braemore Great Bridge described in Appendix A2.4 and Armstrong Bridge in Appendix A2.8.

Examples of lattice details are shown in Figures 4.3 and 5.10.

## Box girders

The Conway and Britannia Bridges are early examples of box girders constructed in riveted wrought iron. The success of these bridges led to a boom in tubular and box girders for the next 20 years but they rapidly dropped out of favour after being criticised by Baker in the 1860s as being "the most unfavourable and uneconomic form possible" (James, 1981).

Box girders came back into favour in the 1950s as welded steel plate assemblages. One of the first and largest box girders constructed in Britain was the aerodynamically shaped deck of the Severn Suspension Bridge, constructed in 1966. At this time box girders were seen as being advantageous through having a clean and elegant appearance not cluttered by stiffeners, as well as being economic. A generation of box girder bridges was constructed including many short-span motorway underbridges and overbridges, for example see Figure 2.10.

**Figure 2.10**     *Steel box girder motorway overbridge*

By this time designers had moved ahead of the contemporary design codes such as BS 153 (BSI, 1968) and some designs turned out to have an inadequate margin of safety in relation to their stability. This came to a head with the near collapse of a bridge across the Danube followed later by collapses and fatalities during construction of the Cleddau in Wales and Yarra in Australia. A short while later, in 1971, a bridge across the Rhine at Koblenz collapsed, also with fatalities. In Britain, these events led to setting up the Merrison Inquiry in 1970 and the subsequent publication of interim design rules for steel box girder bridges, followed some years later by BS 5400. The strengths of existing box girder bridges, including those under construction, were checked and it was found that most required strengthening. Using the new design rules, box girder bridges became less economic than those having plate girders. There were also more comprehensive requirements for health and safety during construction. The interest in steel box girders waned and very few have been constructed since the 1970s.

## Plate girders

Plate girders were introduced in the second half of the 19th century using riveted wrought iron construction as shown in Figure 5.3. Those of the type shown there acted as parapets as well as edge girders. The webs of plate girders usually required vertical stiffeners to resist buckling. In riveted construction these consisted of a flat plate sandwiched between two angle sections, see Figure 4.6. Bearing stiffening at support points was provided by stronger vertical web stiffeners. Additional flexural capacity was provided by cover plates that were riveted to the central portions of the flanges. Depending on requirements multiple cover plates could be used.

In the early 1900s, steel became the preferred construction material with riveting continuing to be used until the mid 1950s.

In the mid 1930s it became possible to weld large assemblages of steel plate and the first welded bridge, on the Billingham Branch Line, was built in 1934. By the 1950s welding had become the preferred method of fabrication. On welded plate girders web stiffeners were composed of flat plate and bearing stiffeners were usually of thicker plate as shown in Figure 2.11, or two closely spaced plates, Figure 10.7.

**Figure 2.11**       *Example of vertical web stiffener*

Transverse stability of the girders is obtained through so called wind bracing or full depth crossbeams.

Due to inadequate knowledge of fatigue performances, some of the welded connections in steel bridges constructed in the period 1960 to 1980 have turned out to be prone to fatigue. Connections that have posed problems include:

- vertical stiffener-to-bottom flange of plate girders
- ends of cover plates welded to flanges of plate girders.

## Through bridges

Through deck construction is common for bridges under and over railways and waterways where minimum depth is required under the running surface. The main longitudinal edge girders are typically plating or trusses made from wrought iron or steel and having crossbeams or trough sections spanning transversely from the bottom flanges. These form half-through decks. Common forms are plate girders restrained by U-frames as shown in Figure 5.3, or truss girders, an example of the latter is shown in Figure 2.12.

Figure 2.12          *Half-through girder rail bridge*

For longer spans, the edge girders are usually trusses and in some cases transverse cross-bracing is fixed between the top flanges to form through decks, see Figure 2.13.

Most of the early truss bridge construction was in wrought iron with riveted connections. Later truss construction was of steel and continued up to the 1930s when it was displaced by welded steel plate construction. More recent trusses have been constructed using fewer and more efficient hollow members.

The multi-modal connections of truss bridges, and particularly the early ones, form dirt traps that lead to corrosion problems.

**Figure 2.13** *Through girder canal bridge*

## Suspended bridges

In the context of this guide suspended bridges are limited to spans less than 50 m and include conventional suspension, cable-stayed and hybrid combinations of the two.

There are a surprising number of short and medium span suspended bridges of all ages in Britain. These range from the first generation bridges of the 1820s and 1830s constructed in iron and timber, to the modern steel bridges. Most of the first generation bridges are protected by having heritage status, being scheduled as national monuments, or listed. Few have survived in their original condition. In some cases problems with wind induced vibrations required strengthening soon after construction, for example Telford's Menai suspension bridge. In the early bridges the main suspension elements were usually chains and the hangers were typically wrought iron rods of circular or rectangular section. Many have required substantial strengthening, load restrictions, or change in use, to enable them to remain in service, for example Union Bridge (Miller, 2006).

Many suspension footbridges were constructed in the period between 1890 and 1930. These were mainly of steel truss construction and many of the later ones have survived in close to their original condition see, for example, Figure 2.14. The main cables and hangers of these shorter span bridges were usually of spirally wound wires protected by galvanizing and paint. This form of construction has generally performed well and required only minimal maintenance work.

Since the 1970s cable-stayed bridges have become more economic to build and very few suspension bridges have been constructed.

**Figure 2.14**    *Suspension footbridge typical of 1930s design*

## 2.3.2    Decks

### Jack arches

Brick jack arches springing transversely from longitudinal iron or steel beams became common practice particularly in the construction of road bridges spanning railways. Typical design features of early jack arches include:

- use of cast iron beams having one or two ring brick jack arches of semi-circular or segmental profile and springing from the bottom flanges of the beams

- fill composed of rubble, concrete or masonry bonded with mortar, see Figure 2.15.

Jack arches springing from wrought iron or steel beams were sometimes sprung from part way up the beams and supported on angles riveted to the webs. Typical aspect ratios for jack arches (beam spacing/rise of the arch) were around 4 and are considered to be non-compliant if greater than 10.

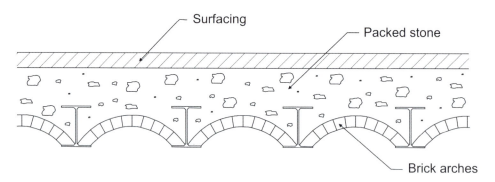

**Figure 2.15**    *Early jack arches*

Lateral restraint to the outer bays of transverse spanning jack arches was usually provided by tie-rods. In practice these can frequently be seriously weakened or effectively lost through corrosion.

Construction of jack arches continued up to c1900.

## Buckle plates

In some cases mass concrete was used for jack arches with permanent metal plate formwork. The plates are known as buckle plates or hogging plates. They were usually fixed to bottom flanges of the beams but not always. When the concrete was filled to the top flanges of the I-beams the construction could be considered as an infill deck as in Figure 2.16.

Buckle plates were made from wrought iron or steel. The plates were usually curved upwards (hogging) but, less commonly curved downwards (sagging), or flat. Plate thicknesses were ¼ in to 1 in (6.3 mm to 12.7 mm) and plate sizes 3 ft to 6 ft (0.91 m to 1.83 m).

The most common form of buckle plate construction is shown in Figure 2.16, see also Appendix A2.11 for example of corroded and repaired buckle plates.

## Infill decks

Infill decks are essentially slabs composed of I-beams in spanwise or transverse directions and infilled with concrete:

- some of the earliest infill decks probably started life as jack arches having cast iron beams infilled with rubble supported on iron plates (buckle plates) fixed to the bottom flanges of the beams. In later years the rubble was removed and replaced with concrete as shown in Figure 2.1.6. This is, in effect, a development of the jack arch

- towards the end of the 19th century a form of slab deck was developed which was composed of steel I-beams infilled with concrete. In some, transverse reinforcement was used, in many there was none, as shown in Figure 2.17.

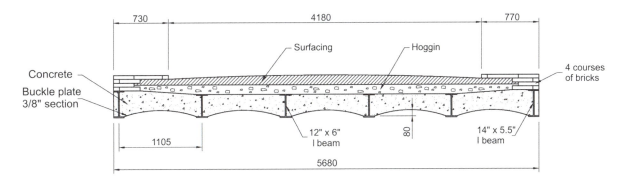

**Note**  Dimensions in mm (except BS beam and buckle plate).

**Figure 2.16**    *Infill deck using buckle plates as development of jack arches*

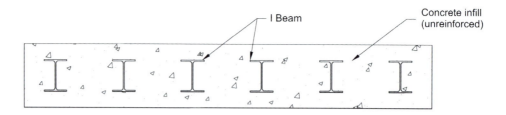

**Figure 2.17**     *Later infill deck*

Many infill decks have been unnecessarily demolished because it has been difficult to demonstrate that they have adequate transverse distribution between beams. However, load tests have demonstrated that there can be adequate distribution through truss action in the concrete. It can also be shown analytically that the strength is often adequate, as described in Chapter 7.

This structural form was popular up to the 1920s for short span bridges and has been used as recently as 2005 (Sreeves, 2007), and is still used in some buildings.

## Trough decks

Trough decks are composed of steel trough-section, shorter span bridges having the troughs aligned in the span-wise direction and longer spans having the troughs transverse to longitudinal girders. The troughs were infilled with rubble, bituminous surfacing material or concrete. The latter may provide composite structural action as a hidden strength but this would not normally be taken into account in first pass assessments.

Construction of a longitudinal trough deck at the turn of the century is shown in Figure 2.18.

**Figure 2.18**     *Construction of a trough deck, 1900*

Trapezoidal troughs were the most common but other patent configurations were manufactured, for example Hobson's section shown in Figure 2.19, Lindsay's patent section and Barlow's cast iron troughs.

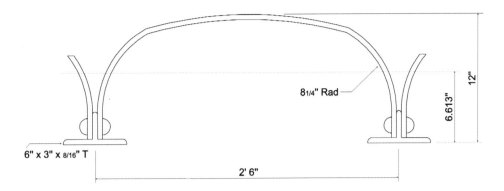

**Figure 2.19**     *Hobson's patent trough*

Transverse connections between troughs could be cover-plated and riveted or lapped and riveted. Cover plates were sometimes riveted to the bottom flanges in order to increase bending stiffness, shown in Figure 2.20.

In one of the few investigations of the structural action of trough decking, a skew bridge (skew span 9 m, square span 7 m) constructed in 1902, was load-tested by two articulated vehicles having nearly identical 2-axle bogies, each loaded to 170 kN. It was found that maximum values of strain measured in the troughs were a factor of three less than calculated values. It was concluded that the hidden strength was due to a combination of end-fixity at the abutments, effect of skew and composite action (Beales and Daly, 1991).

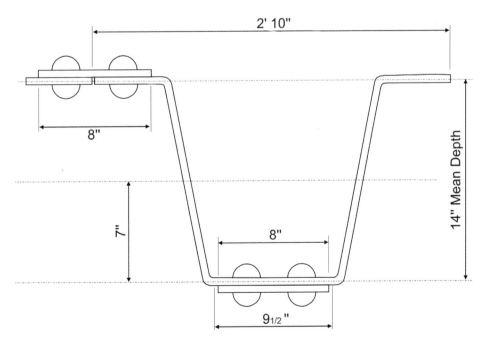

**Figure 2.20**     *Connections between trapezoidal troughs*

Troughing was sometimes constructed from riveted steel sections as in Highbury Corner Bridge, North London, and illustrated in Figure 2.21. This form of construction has the drawback that the vulnerable riveted connections are virtually impossible to inspect.

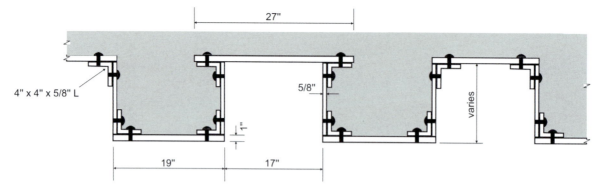

**Figure 2.21**    *Steel troughing constructed from riveted sections*

## Longitudinal timber troughs

On railway bridges longitudinal timber troughs are often present. In this form of construction the deck is usually supported on cross-girders fixed between the main longitudinal girders. The deck plate is recessed to accommodate longitudinal timbers which support the rails. A representative cross-section is shown in Figure 2.22.

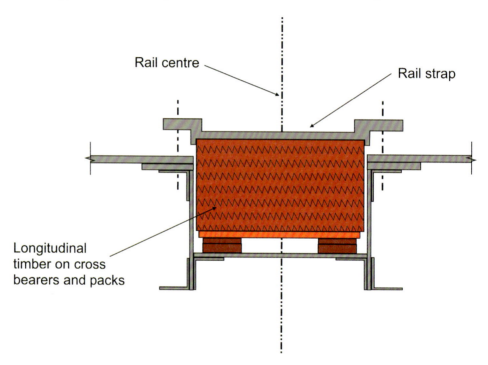

## Cross-section through element of trough

**Figure 2.22**    *Element of timber trough deck*

In Figure 2.23 a different example is shown where the timbers are braced against the inside faces of main girders by timber transoms. Beneath these timbers is solid bitumen bedding of about 40 mm depth. On this bridge the transoms had deteriorated and it was necessary to replace them with steel sections.

**Figure 2.23**     *Example of deteriorated timber*

## Orthotropic decks

Welded steel orthotropic (orthogonal isotropic) decks were introduced in the 1960s. Their main attraction was the light weight that can be obtained from this form of construction. This feature is of advantage to moveable and long span bridges.

An orthotropic deck comprises a deck plate having longitudinal stiffeners and crossbeams welded to its underside, as shown in Figure 2.24. Various types of longitudinal stiffener have been used including bulb flats, V-stiffeners and trapezoidal-stiffeners. The longitudinal stiffeners are usually passed through the crossbeams and cope holes provided to enable longitudinal welds to pass through uninterrupted and avoid having the triple connection between the stiffener, crossbeam and deck. Details of the welded connections to the deck plate are very important to the fatigue performance of the assemblage.

In the UK the running surface is commonly 38 mm thick mastic asphalt, usually laid by hand to cope with cusping in the deck plate that results from the welded crossbeams. In practice the surfacing acts compositely with the deck plate to provide a hidden contribution to its strength. This action is, however, unreliable as it is influenced by temperature and becomes less effective in hot weather. It is also affected by time dependent processes as the surfacing ages and becomes cracked.

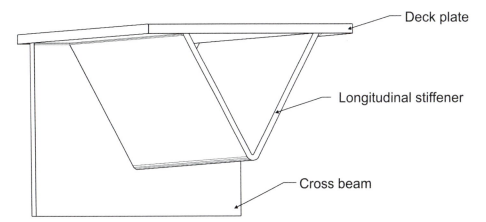

Deck plate

Longitudinal stiffener

Cross beam

**Figure 2.24**     *Outline of welded connection between longitudinal stiffener, crossbeam and deck plate*

For more details of the fatigue performances of orthotropic decks see Gurney (1992).

# 3 Guidance on statutory and other requirements

This chapter deals with statutory obligations in relation to health and safety, heritage, conservation and ecology.

It is important to be aware that legislation is constantly being added and updated. The following is a synopsis of current issues, but should not be regarded as a substitute for up-to-date awareness in each area.

## 3.1 INTRODUCTION

This chapter outlines the statutory obligations of health and safety regulations, statutory undertakers, heritage and environment.

The reader is encouraged to check the latest position with regard to legislation as amendments are introduced from time to time which may prove to be more onerous than the current situation outlined here.

## 3.2 HEALTH AND SAFETY

### 3.2.1 General

The Health and Safety at Work Act of 1974 provides the framework for safe working. The principal regulations under this Act that are relevant to the maintenance of iron and steel bridges are:

1   The Management of Health and Safety at Work Regulations of 1999. The most important requirement of these regulations is for employers to assess the risks to employees and others, and take steps to eliminate or reduce these risks. Requirements also include co-operation between employers on health and safety matters and the provision of information.

2   The Construction Design and Management Regulations (CDM) 1994 (as amended in 2007), require all those involved in designing and planning construction work to consider the health and safety risks that may arise during construction so that these risks can be eliminated or reduced. The CDM Regulations require preparation of a health and safety plan, and employment of competent personnel. The scope of the regulations is wide covering new works, remedial works and certain examinations of structures.

3   The Construction (Health, Safety and Welfare) Regulations of 1996 (currently under revision), set the minimum standard for conditions on a construction site including protection from physical hazards and provision of welfare facilities.

### 3.2.2 Construction operations

There are a number of other items of health and safety legislation which are relevant to working on iron and steel bridges, including:

1   The Control of Substances Hazardous to Health (COSHH) Regulations 2002, covers the control of risks to employees and other people arising from exposure to

harmful substances generated through any work under the employer's control. The Regulations set out a simple framework for controlling hazardous substances in the workplace. Employers must ensure that the exposure of employees to hazardous substances is either prevented or adequately controlled. These substances may be included in the fabric of the bridge or used in carrying out the works. Materials used in bridge repair, such as resins, may be hazardous in their application, and work which generates dust and fumes may present a hazard to those carrying out the works and to members of the public. Materials in an existing structure may include potential pollutants or those harmful to health such as lead and asbestos.

2   Health hazards may also arise from wildlife, for example pigeon droppings, rat infestation, bat infections etc.

3   The Control of Lead at Work (CLAW) Regulations of 2002 apply to any type of work activity such as handling, processing, repair, maintenance, storage, disposal etc which is liable to expose employees and any other personnel to lead. The regulations place a duty on employers to prevent or adequately control the exposure of employees to lead irrespective of the source of that exposure. The exposure to lead may result from work with lead or lead compounds being carried out by the employer's own employees, or incidental exposure arising from work nearby being carried out with lead compounds by another contractor working on the site. Exposure can occur when old original lead-based paint is being cleaned from old bridges during maintenance painting, see also Chapter 11.

4   The Environmental Protection (Controls on Injurious Substances) Regulations of 1992 control the marketing and use of paint containing white lead but allow restricted use of lead paint in accordance with the European Marketing and Use Directive (89/677/EEC) in strictly controlled and special circumstances for the refurbishment of historic structures, see Section 3.4. Specialist advice should be sought for such work.

5   Lead wastes should be disposed of in accordance with the Environmental Protection (Duty of Care) Regulations 1992. It may be necessary to check with the local environmental health officer and/or waste regulatory authority for any special disposal provision.

6   The Working at Height Regulations (WAHR) of 2005 apply to all work at height where there is a risk of a fall liable to cause personal injury. They place duties on employers, the self-employed, and any person who controls the work of others such as bridge managers or owners who may contract others to work at height. The inspection of bridges almost always involves working at height, sometimes over rivers or other hazards and it is essential that anyone planning an inspection is familiar with the key points of the regulations. Information can be obtained in the HSE's brief guide to the regulations (Health and Safety Executive, 2005a).

7   The Confined Spaces Regulations of 1997. A confined space is any enclosure, such as a steel box girder, where there is a potential risk of serious injury from hazardous substances or dangerous conditions. A booklet providing information and advice for people involved in this type of work has been published by the HSE (Health and Safety Executive, 2005b).

8   The Provision and Use of Work Equipment Regulations of 1998 includes supervision and approval of trained personnel, and prevention of access and misuse by unauthorised persons. As with any other civil construction work, there are risks associated with the use of heavy plant, such as cranes and excavators, light plant, such as generators, and hand-held tools, such as angle-grinders. Special access equipment is often required which can also bring hazards to the work.

9   Lifting Operations and Lifting Equipment Regulations (LOLER) of 1998. The Regulations aim to reduce risks to the health and safety of operatives and others from lifting equipment provided for use at work. Generally, the Regulations require that lifting equipment provided for use at work is:

    o   strong and stable enough for the particular use and marked to indicate safe working loads

    o   positioned and installed to minimise any risks

    o   used safely, ie the work is planned, organised and performed by competent people

    o   subject to ongoing thorough examination and, where appropriate, inspection by competent people.

10  The Noise at Work Regulations (2005) require employers to:

    o   assess the risks to employees from noise at work

    o   take action to reduce the noise exposure that produces those risks

    o   provide employees with hearing protection if noise exposure cannot be reduced enough by using other methods

    o   make sure the legal limits on noise exposure are not exceeded

    o   provide employees with information, instruction and training

    o   carry out health surveillance where there is a risk to health.

11  There are no regulations specific to welding operations on steel bridges. However, there are legal duties applicable under the more general health and safety legislation described above.

A more comprehensive description of current health and safety legislation can be found in Tyler and Lamont (2005).

## 3.3    STATUTORY UNDERTAKERS

Statutory undertakers are organisations licensed by the government to dig holes in the roads, verges, footways (pavements) under the New Roads and Street Works Act (NRSWA) 1991. They include all utilities electricity, gas, water, telephone, cable telephone and television, and other telecommunication companies. All works must be carried out by qualified people holding an NRSWA certificate. Bridges commonly carry utilities and can be subject to the requirements of NRSWA. On some of the older bridges, the location and types of utilities being carried are not always known and it is necessary to excavate trial holes in the deck.

Statutory undertakers are regulated by codes of practice which cover the following:

* reinstatement of openings in highways

* co-ordination of street works and works for road purposes and related matters

* measures necessary where equipment is affected by major works

* works inspections.

The activities of statutory undertakers are normally controlled by local authorities who will carry out inspections of the works at various stages of construction.

## 3.4 HERITAGE AND CONSERVATION

To varying degrees, all iron and early steel bridges are of historic significance as they are records of the industrial age. In consequence the bridges rated as most important are subject to statutory controls; they can be scheduled under the Ancient Monuments and Archaeological Areas Act (1979) or listed under the Planning (Listed Buildings and Conservation Areas) Act 1990. In addition to scheduling and listing, bridges may be afforded protection under a variety of designations of the land on which they are sited, for instance in a conservation area, a site of special scientific interest (SSSI), or a special area for conservation (SAC).

Non-statutory planning policy guidance notes (PPGs) provide guidance to local authorities and others on planning policy and the operation of the planning system. Those most relevant to the management of iron and steel bridges are Planning Policy Guidance 15 (1994) and Planning Policy Guidance 16 (1990).

Works on listed or scheduled iron and steel bridges, and those within areas having special environmental protection, require consultation and cooperation with the local authority planning department and relevant heritage body:

- English Heritage.
- Historic Scotland.
- Environment and Heritage Services, Northern Ireland.
- Cadw Welsh Historic Monuments.

Further discussion of legislation across the UK concerning historic bridges and guidance on issues associated with planning and permissions is included in the publication *Conservation of bridges* (Tilly, 2002). Guidance on sympathetic alteration is given in the Highways Agency publication *The appearance of bridges and other highway structures* (HA, 1996) and *How to handle scheduled and listed structures* (Streeten, 1990).

## 3.5 ENVIRONMENT AND ECOLOGY

The Wildlife and Countryside Act of 1981 and the Countryside Rights of Way Act of 2000 provide the principal legislation for protection of the environment and ecology.

Conservation bodies having responsibility for promoting the conservation of wildlife and natural features in the UK are:

- Natural England, (formerly English Nature)
- Scottish Natural Heritage
- Northern Ireland Environment and Heritage Service
- Countryside Council for Wales.

Factors that should be considered in relation to environment and ecology include:

- management of environmental impact
- use of new materials
- re-use and recycling of waste
- pollution
- wildlife.

Under the Habitat Regulations of 1994, operations on SAC's or SSSI's can be banned if it is considered that the work could cause environmental damage.

Useful advice on dealing with problems of environment and ecology is given in publications on *Working with wildlife* (Newton *et al*, 2004), *Environmental good practice on site* (Coventry and Woolveridge, 1999), *Urban environment and wildlife law* (Rees, 2002), and *Environmental sustainability in bridge management* (Steele, 2004).

Advice is given on waste minimisation in construction by Addis and Talbot (2001) and reclaimed and recycled construction materials by Coventry *et al* (1999).

# 4    Asset management and preventative maintenance

> This chapter gives an overview of the principles of asset management, their influence on condition appraisal and the choice of remedial treatment.
>
> An introduction is given to the principles of preventative maintenance.
>
> Examples are provided of typical features that have led to the need for remedial work.

## 4.1    ASSET MANAGEMENT

### 4.1.1    General

A condition appraisal of a metallic structure should consider the particular circumstances and attributes of that structure to enable a strategy for maintenance or repair to be defined.

The context within which the maintenance or repair activity will be undertaken will have a significant impact on the decisions, and conclusions to be drawn. In most cases, the structure in question will be a part of a larger bridge stock. On occasions when there is doubt, the ownership of the structure may have to be confirmed. It will be necessary to apply informed judgement to ensure that available financial and labour resources are used to best effect.

Over the last 15 years, systems have been developed to assist with this task, and research is ongoing within the UK and internationally to improve the processes and facilities that are available.

Difficulty (and expense) of access is often cited as a reason for poor maintenance. A well-structured system for asset management will take this into account, plan for it, and provide the necessary evidence to support application for the required funds. The system should also be able to demonstrate the long-term financial consequences of not carrying out timely maintenance (see Section 4.1.1).

In this chapter, a brief overview of bridge management is given, to assist the reader in understanding how decisions on an individual structure may be influenced by the management regime in place. Suggestions for further reading are given for those wishing to research the subject of bridge management in more depth.

It should be noted that the term bridge management system is often used to refer to software produced commercially to assist the bridge management engineer in handling structure data and the bridge management process. Within this document, the term (abbreviated to BMS) will refer to the whole system of managing a bridgestock, including processes and procedures, databases, work scheduling and computer systems etc. The influence of the BMS on an individual structure will be considered and specific instances related to metallic structures described.

## 4.1.2      The basis of asset maintenance

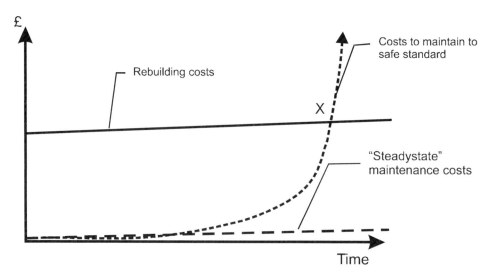

**Figure 4.1**      *Indicative maintenance cost curve*

Figure 4.1 shows an indicative maintenance cost curve. This curve can be applied equally to an element of a structure, a structure, or a group of structures. The explanation below is given in terms of a single structure.

If the structure is properly managed from new, with small but consistent funding provided to undertake regular inspection and routine maintenance, continued good performance may be expected, shown by the dashed line in the diagram. The increase in slope of the line is indicative of the need to allow for the effects of cost inflation.

If maintenance of a structure is ignored following construction, despite lower financial outgoings, long-term deterioration processes will be ongoing. Eventually the lack of attention to routine maintenance will lead to acceleration in the processes of deterioration, and significant escalation in the costs of restoring the structure to an acceptable condition. This is shown by the dotted line in Figure 4.1.

At some point (shown as X), the costs of repairing the structure will equal or overtake the costs of complete replacement. Note that this illustration presumes that the structure has an adequate (but deteriorating) margin of safety until point X is reached. In some instances there may be a need for earlier intervention on the grounds of safety of a critical element.

The hidden costs of traffic congestion can have a major influence on whether or not to replace a structure and it may be necessary to carry out a more expensive refurbishment, or to replace early in anticipation of traffic growth.

In practice the continuous curves shown in Figure 4.1 may be more saw-toothed, reflecting a number of small maintenance interventions over a period of time.

## 4.1.3      The aims of a bridge management system

A bridge management system comprises a collection of processes, procedures, human and IT resources established with the aim of:

- regular collection and recording of data relating to the condition and assessed capacity of structures, or elements of structures, in a consistent manner

- processing the data to determine priorities for maintenance activity and expenditure

- predicting rates of deterioration in such a way that various maintenance strategies can be assessed

- producing cost estimates for future financial planning

- planning and organising maintenance and repair activity

- issuing and tracking orders for work to be carried out

- recording the benefits of the work such as improvement to the structures

- reporting on the efficiency of spending funds, so that best value can be demonstrated

- retaining, updating and maintaining records in a format that is readily retrievable (note that electronic storage is not always robust as storage formats become outdated).

Activities within a BMS are usually budgeted under a number of headings. A typical division would be:

- routine maintenance works (gulley cleaning, vegetation removal etc)

- minor works (minor repairs, touch-up paintwork)

- capital works (major repairs, reconstruction)

- special projects

- emergency works.

The operation of the BMS (and the health of the bridgestock, and efficiency of managing it) is measured using a number of performance indicators. These vary from system to system, but typically would include:

- condition of the bridgestock

- reliability (in terms of safety)

- availability (bridges with speed or weight restrictions)

- structures workbank (the financial value of outstanding work required to bring all the structures up to a desirable standard as defined by the UK Bridges Board).

The principles behind a BMS have been established for some time. The operational aspects of inspection of structures, recording the data, undertaking some form of prioritisation, and issuing work orders have also become well-established. Development of bridge management systems now focuses on improving the accuracy with which they can be used to predict future deterioration of elements of a structure, a structure, or the bridgestock as a whole. In this way, future maintenance trends can be identified, and the impact of various funding levels and strategies investigated.

Typical of the challenges to be faced include different bridge forms, variation of the operating conditions of a bridge or structural element, and lack of robust data on deterioration of structural elements

Improvement in techniques for dealing with these variations will lead to enhancement in the use of the BMS. Even with the current generation of BMSs, the approach is extremely useful, and better prediction will be possible as more research data are gathered. Ultimately, a BMS can be used to determine and to justify the funding that is needed to maintain the bridgestock, and to predict the possible knock-on effects of underfunding. The BMS will support making best use of funds that are provided, and to demonstrate that best value has been achieved

Most major bridge owners have established a BMS, usually tailored to their specific requirements. Developments within the area of bridge management are reported in the technical press, and symposia such as those at University of Surrey, Harding *et al* (1993, 1996 and 2000), and Parke and Disney (2005). Work by the UK Bridges Board has resulted in a comprehensive document aimed at BMS for Local Authorities (UK Bridges Board, 2005). This document contains material related to planning and management of bridge stocks, and provides a good basis for understanding the process.

## 4.1.4 An asset management framework

The UKBB code document proposes that asset management can be broadly categorised at three levels:

1   Strategic – where are we going and why.

2   Tactical – what is worth doing and when.

3   Operational – how to do the right things.

A fully functional BMS will deal with the large volume of data to be handled at the operational level, and process the various indicators and trends to advise decisions at the tactical and strategic levels. The UK Bridges Board code gives a detailed description of the process for prioritising work and making best use of the funds that are available.

Ultimately, a bridge owner is faced with making a decision on how the health of the bridgestock is to be managed. Broadly, these are:

(a)   To maintain the condition of the bridgestock at current levels of performance.

(b)   To improve the overall condition of the bridgestock.

(c)   To manage the bridgestock towards eventual (planned) replacement through minimal intervention.

Each of these strategies can be reflected at the level of an individual structure. Moreover each will require a different level of funding input, which has to be justified to a financing department in competition against other demands, often outside the transport sector. In practice, most bridge owners are faced with strategies somewhere between options (a) and (c) above, and have to balance requirements against the limited funds available. Within this context it is necessary to define a management regime advised by the overall maintenance strategy for the bridgestock, and the condition of the structure or element.

In considering the impact of structural works, it is important to recognise that the general public is the ultimate client, and that their perceptions, priorities and concerns may be different from those of the bridge owner. Public concerns may arise from (typically) serviceability failures that may not be significant in themselves, but should be investigated, as they may indicate other problems. Typical examples could be:

*   damage to fixtures and fittings on the bridges

*   loose concrete

*   flaking paint

- structure groaning due to sticking bearings or joints

- historical, cultural, or in some cases, sentimental aspects of the bridge

- vibration or deflection perceived to be excessive.

Excessive vibration or deflection should always be investigated as the movements can lead to fatigue damage. Footbridges that demonstrate such behaviour are likely to attract unwanted attention from vandals trying to make the bridge bounce and, while this is unlikely to bother the more robust members of the public, elderly and infirm people are fearful of the movements and are likely to avoid using the footbridge.

## 4.2 PREVENTATIVE MAINTENANCE

### 4.2.1 Issues of concern to bridge owners

When establishing a bridge management regime, or a maintenance intervention on a particular structure, the following issues are of importance to the bridge owner:

- safety of the structure and those that may be affected by it

- understanding the current condition of the structure.

- assessed load capacity of the structure (likely to be less than the true capacity)

- potential to accommodate increased loading (eg extra traffic lanes)

- the residual life of the structure (if known), set against the required future life

- availability of the structure (ie whether a load restriction is required)

- reliability of the structure (ie the likelihood of load restriction or damage)

- whether or not the structure is fit for the purpose that it is needed

- available budget for maintenance or repair

- the security of cost prediction (how accurately can the costs of a maintenance operation be forecast)

- public perception of the state of the structure (eg the structure may have large deflection, or peeling paint, but in reality is structurally sound)

- the means and costs of providing access to the structure for inspection and maintenance.

Operation of a BMS aims to provide a means of managing these issues, and to support decisions with robust data and procedures for handling these data.

When assuming responsibility for a bridgestock, the bridge manager should confirm the current condition from existing records, and address, as a matter of urgency, any items that are safety-critical by means of special projects. If not already in place, a system of routine maintenance and minor works for non-critical structures should be installed.

When managing issues of concern, it is important to ensure that secondary members are not overlooked. For example, failed wind-bracing falling from the superstructure could lead to accidents on roads or railways beneath.

Over a period of years, bridges should be maintained to a level where performance meets required criteria. In parallel with this, a programme of preventative maintenance is needed to ensure that this level of performance is sustained. Having achieved this steady state condition, a strategic decision (see Section 4.1.4) may be made to provide funds to improve the overall condition of the bridgestock.

## 4.2.2      The condition of a structure

### Establishing the condition

The condition of a structure is usually identified by a combination of inspection, site measurement, reference to existing records, and structural assessment to determine serviceability and load-carrying capacity. Regular observation is needed to obtain information regarding the rate of deterioration of the various elements of the structure.

An analysis of the structure (which should not be complex) will show which elements are critical to its load-carrying capacity and those which may be subject to deterioration over time from corrosion and fatigue.

Structures are usually inspected under a regular regime determined by the bridge owner, see Chapter 6 for more information. Suggestions are made by the UK Bridges Board (2005), but typically, such a regime would include:

- a principal or detailed inspection every four to six years, where the structure is closely examined. The criterion frequently quoted is "within one metre", or "within touching distance", but this is not always possible in practice

- a general or visual inspection every one or two years, between the detailed inspections

- special inspections for structures where there is concern, for example, a quarterly or monthly walk through inspection

- provision for emergency inspection in response to an incident (bridge strike, or inundation).

Increasingly, it has been recognised that a one size fits all inspection regime is not always necessary ; the time between inspections can be lengthened for bridges of simpler, more robust forms of construction, that are in good condition. Conversely, the intervals between inspections can be reduced for structures where it has been assessed that there is a higher risk of deterioration.

The Highways Agency document BD 79/06 (2006) gives guidance on the actions to be considered if a bridge is assessed to be substandard. While intended for use with Highways Agency bridges, the principles within BD 79/06 are suitable for wider application if agreed with the client.

Structures that include moving components such as moveable bridges, and roll-on, roll-off ferry linkspans, will have mechanical components having specific maintenance needs for lubrication, electrical and hydraulic testing. The special nature of these structures should be recognised at the outset. Problems can arise from:

- change of loading condition as the structure moves from the *closed* to the *open* position

- wear of moving parts

- time-related deterioration of seals, hydraulic pipes etc

- infrequent use of the machinery leading to poor lubrication

- lack of regular trial operation

- lack of regular flushing of lubrication

- overloading from berthing or docking manoeuvres.

For elements that require regular monitoring, the ready availability of inexpensive IT equipment that can be discretely sited on a structure enables readings from a range of sensors which can be taken remotely and fed back to a centrally located computer.

## 4.3 KEY ISSUES IN PREVENTATIVE MAINTENANCE

### 4.3.1 General

Preventative maintenance of iron and steel bridges can be considered under the following key areas:

- water management

- damage to protective treatment

- corrosion

- fatigue

- wildlife

- vandalism

- unwanted structural restraint.

### 4.3.2 Water management

Poor management of rainwater falling onto bridge structures and leaking through joints is a major cause of deterioration. Problems can arise when detailing of the drainage system is not properly thought out during design, or if good drainage systems are not regularly cleaned and maintained, leading to blockage and overflow. The problem is compounded during winter months when water runoff from a road deck often contains a high proportion of de-icing salt which, if not properly handled, leads to acceleration of the normal processes of coating break-down and corrosion.

Expansion joints are subjected to heavy wear and tear as a result of traffic loads, and also the longitudinal movements and forces applied during thermal cycles of expansion and contraction. The expansion joints need to be specified, detailed and installed with care, and should be compatible with the selected waterproofing system. A common source of leakage is a defective or unsuitable joint, or poorly executed patch repairs using unsuitable materials.

Suggested measures in relation to water management include:

- drainage paths should be well-defined when the structure is initially conceived, or when remedial works are being designed. The consequences of blockage should be considered

- the ease of access for routine maintenance should be considered

- water runoff and air circulation to all wetted surface should be checked

- all expansion and movement joints should be regularly checked for leakage and water retention. If possible, inspection is best done during, or just after, wet weather

- details that have re-entrant corners, or where debris can collect and trap moisture, should be regularly cleaned

- bridges that are constructed from weathering steel need particular consideration to avoid problems with staining and runoff. Guidance is available in the publication on weathering steel by Corus (2004), see also Section 11.6

- additional protection may be necessary for exposed top flanges of girders, particularly if the protective coating has begun to break down. The action of rain washing away the corrosion products and exposing new metal can accelerate material loss (see Figure 4.5)

- box section members or box girders may suffer from water collection due to air ingress and subsequent condensation within the box. Measures to deal with this moisture by either drainage or ventilation should be considered at the design phase. Note that complete sealing of a large welded box girder is often not possible, and it is preferable to provide ventilation

- details such as buckle plates on decks may have sagged either during construction or owing to subsequent deterioration. This action produces a potential water trap and remedial measures will need to consider additional drainage.

Two examples of deterioration arising from water management issues are given in Boxes 4.1 and 4.2.

**Box 4.1**    *Water management of a bascule bridge*

This example is included to show how the lack of inspection and maintenance to a simple drainage detail can lead to situations that are potentially dangerous to operating and inspection staff, and can result in expensive repairs. The specialised nature of the structure is incidental, although the presence of deep chambers, and electrical and mechanical equipment could have significantly exacerbated the damage and danger to staff.

A bascule bridge, constructed in 1941, spanned 20 m across a harbour entrance. The single-leaf, single-span structure is supported on one side on a substantial hollow pier structure, having deep basements over two full storey-height levels. The basements were used for equipment storage, and were connected by steel ladders. The arrangement is show in Figure 4.2

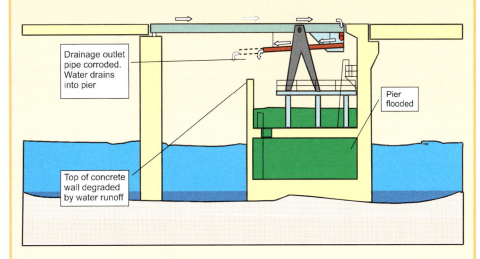

Figure 4.2        *Damage to a bascule bridge caused by water mismanagement*

The drainage for the bridge deck was arranged to fall into a collector pipe at the counterweight end of the bascule, where an outlet pipe took the water back along the underside of the deck, and out into the harbour. The pipe was of ductile iron.

Over the years, the end of the pipe had rusted away, and shortened the outlet. Initially all the runoff splashed onto the concrete wall of the pier, causing deterioration of the reinforced concrete parapet wall. Further shortening of the pipe resulted in the entire runoff being diverted to fill the basements of the pier and flooding the materials stored there. When inspected, two basements were found to be completely flooded. The water and corroded manhole covers in the machinery floor presented significant hazards to inspecting engineers as well as causing long-term damage to the structure.

Regular inspection of the drainage, and installation of a 3 m length of plastic drainage pipe would have prevented the flooding, and the associated deterioration of the parapet wall and basements. This example is based on a bridge which has since been repaired.

**Box 4.2**    *Water management of a lattice girder bridge*

This example illustrates the difficulty in designing a corrosion-free detail where a lattice girder meets the roadway.

Figure 4.3 is of a lattice girder bridge constructed in the late 19th century. A major refurbishment was carried out c1990. Lower parts of the lattice which had corroded were replaced by steel plate, and a deeper road construction was added to spread local wheel loads more effectively.

As shown in the part cross-section below, channels were formed at either side of the carriageway to direct the water to drains at the end of the bridge. The channels were U-shaped to direct water towards the centre of the channel, and away from the wrought iron trusses at the edge of the road.

The gap between the crossed-members of the lattice is shown in Figure 4.3. This gap is of the order of 10 mm, and the dimension is dictated by the connection plates forming the webs of the T-shaped flange members at the top and bottom of the girder. A combination of the original design and the refurbishment needs made the detail difficult to avoid and it will not be easy to ensure that proper protective treatment is applied to the space between the two members. A better solution where possible is to use either a wider gap, or to provide a larger filler piece that can be more easily painted.

Figure 4.4 shows that despite the U-channel and longitudinal slope, water can still pond against members of the lattice. Poor shaping and a relatively small amount of leaf debris are contributing to the ponding. Already there is evidence that local breakdown of the protective treatment has occurred, and corrosion has started again.

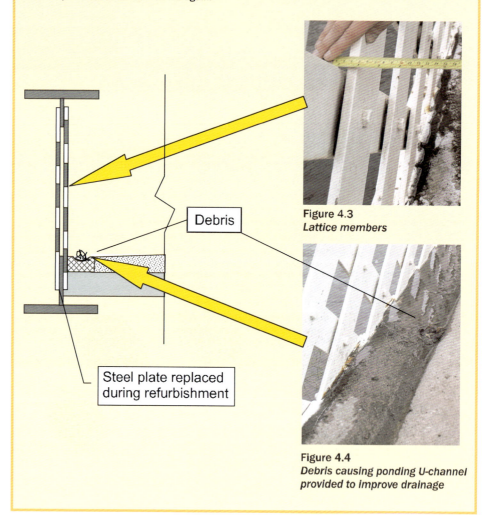

Debris

Steel plate replaced during refurbishment

Figure 4.3
*Lattice members*

Figure 4.4
*Debris causing ponding U-channel provided to improve drainage*

### 4.3.3    Damage to protective treatment

Protective treatment is dealt with in more detail in Chapter 11 and comment in this section is limited to general issues. Initial breakdown of a protective coating should always be monitored, and patch repairs considered, although in some instances it may be more cost-effective to delay intervention to address a larger area of repair in a single contract. When planning an intervention it is essential to identify the root cause of the problem and take appropriate action.

The operating conditions under which a coating is expected to perform can radically affect its life. In North America an extended life for protective coatings on major structures is achieved when regular washing or steam cleaning of the surface is undertaken, particularly in splash zones affected by salt spray.

It is important to recognise that not all coatings can be easily repaired, however effective they are as corrosion prevention. In particular, galvanizing or hot metal spray coatings can be difficult to patch repair after they have been initially applied, due to oxides forming on the surface of the metal coat. The oxides do not always bond well with material applied some time after the initial application. If in doubt, the advice of a coatings specialist should be sought for the particular protective system being considered.

When examining the protective treatment on a structure, it is important to recognise that defective coats of paint may have been over-coated. This can hide structural problems such as rivet heads which are present, but broken and structurally ineffective. It should be noted that successive applications of over-coated paint can add significantly to the deadweight of the structure.

### 4.3.4    Corrosion

Corrosion of structural steel presents problems as the rust formed does not adhere strongly to the surface of the metal. The rust may flake off, or be removed by rain or wind, exposing fresh metal to further corrosion. An example of this is shown in Figure 4.5.

**Note**    This viaduct was constructed in the late 19th century and has been disused since the mid 1960s. The top flanges of the cross-girders have been subject to direct exposure to the rain, and corrosion products are regularly washed away, resulting in much higher loss of section compared to the bottom flanges of the same members.

**Figure 4.5**    *Corroded girders*

When they are not removed, the corrosion products are porous and absorbent so that they can retain water local to the detail and encourage further deterioration. Where the timber deck is in place, and drainage is not properly managed, the supporting steel can suffer extreme local corrosion and in one instance failed altogether so that the deck fell through the superstructure.

Rust formed during corrosion is greater in volume than the parent metal, and the high levels of expansive forces generated between the plates of a connection can lead to bending the plates or bursting the rivets.

Corrosion of iron and steel structures can be prevented in a number of ways. Most rely on either protective treatment of the material to exclude the elements needed for corrosion (oxygen and water), or by controlling the electrolytic processes by use of more reactive metals or by imposing an electric potential.

Carbon steel can be electrically protected by coating with a less noble metal such as zinc (eg galvanizing). In addition to having a somewhat tougher oxide, the zinc corrodes in preference to the steel.

Corrosion of immersed steel structures can be protected by cathodic protection. There are two forms, impressed current and sacrificial anodes.

In impressed current systems an electric potential is applied between zinc anodes and the structure. The steelwork is protected as the (sacrificial) zinc anodes corrode in preference to the structure.

Where the structure is submerged, sacrificial anodes comprising blocks of zinc are electrically connected to the structure to form a large corrosion cell. The cell is driven by the potential difference between the zinc and steel.

Cathodic protection may be a useful tool in retrofit situations but these techniques require specialist advice.

## 4.3.5     Fatigue

Fatigue damage to an element of a structure is caused by repeated loading, leading to crack initiation and propagation, and to eventual failure of the element when the crack has grown to the extent that the remaining section can no longer carry the loads to which it is subjected. Information about detection and monitoring of fatigue cracking is given in Chapter 6. **The engineer should be certain that brittle failure of the material is not an issue before moving on to consider fatigue-related failure.** Guidance on the key features that affect fatigue is given in the Eurocode for structural steelwork, BS EN 1993-1-9 (BSI, 2005) and in BS 5400-10 (BSI, 1980).

Older structures may not have been designed with fatigue as a consideration and may require special attention. The materials from which they are constructed are metallurgically less clean than modern steels, and so may give rise to damage in locations around connections (bolted, riveted or welded).

Rivet and bolt holes may not have been well formed, and areas around these connections may be potential sites for crack initiation. Typical welded and bolted details are shown in the fatigue codes referred to above, with an indication of the relative performance of the detail type.

In dealing with a suspected fatigue problem, expert advice should be sought to confirm initial findings. It is sometimes possible to justify a do-nothing option, providing that a regular maintenance regime is in place for the structure. An appropriate strategy may be through monitoring accompanied by a contingency plan.

Structures that are fatigue-sensitive can have their life span extended by reduction of the frequency or magnitude of load effects that are experienced by the critical elements. In some cases, a more detailed structural analysis can give better information on load effects, and lead to increased confidence in the detail.

Material can be added to the structure local to the detail to reduce the calculated stress range. This strategy needs to be carefully considered – sometimes attempts to strengthen a fatigue-prone detail by adding material may result in additional stresses due to welding, or damage during preparation.

An alternative strategy is to provide additional bracing or stiffening to the structure so that the revised load paths are arranged to divert load away from the critical details. How this was achieved during the upgrade of the Docklands Light Railway viaducts in London is described in the papers by Gonsalves and Deacon (1990), and Pilgrim and Pritchard (1990).

## 4.3.6     Damage caused by wildlife

The most common damage caused by wildlife is by their excretion and particularly the guano deposited by pigeons. The installation of netting is a preventative action commonly taken, as illustrated in Figure 5.13, but this requires regular maintenance. Pigeons have a strong homing instinct and once they adopt a roost it is difficult to deter them from continuing to use it. To avoid the problem, it is often necessary (but not commonly known) to cull the pigeons that have made the bridge their home before installing the netting.

A recently developed preventative action is the installation of stainless steel mesh which can be fixed to bottom flanges to prevent access to the favoured roosting places, an example is shown in Figure 4.6. The mesh is designed and fitted so that it can easily be removed to enable maintenance such as painting to be carried out. For further details see, for example, <http://www.glideholdings.com/>.

**Figure 4.6**     *Prevention of access by pigeons (the photograph shows Pigeonglide™)*

## 4.3.7 Damage caused by vandalism

Preventative actions to avoid vandalism include regular maintenance to remove fly-tipped combustible material such as used pallets and old car tyres. It is sometimes necessary to erect fencing to prevent access to vulnerable parts of a bridge. Unfortunately vandals, like pigeons, are difficult to deter once they have taken an interest in a bridge.

Graffiti, although unsightly, is not damaging but can involve offensive or obscene messages which have to be removed without undue delay.

## 4.3.8 Unwanted structural restraint

An important cause of deterioration of iron and steel bridges is through unwanted or unintended restraint. Restraint of this nature can occur at bearings, expansion joints, or due to overall (large scale) lack of articulation of the structure. Although these effects can be present in any bridge, they can be more serious in metal structures, as large horizontal restraint forces can be generated very easily. These forces can exceed those due to braking or skidding for which bridges are designed, and can lead to buckling of slender components or failure of the bearing arrangements.

The effects of climate change may mean that the operating thermal range experienced in practice differs from that assumed in the original design (it has been calculated that climate change could cause increases in maximum return temperatures of 0.5 to 1.0°C). It is important to ensure that sliding bearings and joints are able to utilise the full extent of their design movements without causing restraint.

**Box 4.3**    *Bearing restraint forces*

Boston Manor Flyover is a three-span viaduct (64.5 m, 111.25 m, 64.5 m) carrying the M4 motorway west of London, see Figure 4.7. The 18.3 m wide roadway is carried on a composite steel and concrete deck that spans transversely onto two steel trusses, spaced 11.3 m apart. The trusses vary in depth from 2.7 to 9.5 m over the piers. The central 36.6 m of the main span is supported on four half-joints, each with a fabricated sliding bearing. The ends of the main span are fixed to substantial piers.

This is an example of how high friction or restraint can lead to damage. Components may be observed to be behaving as intended, and the existence of restraint is only seen when damage occurs.

Figure 4.7    *Boston Manor Flyover, West London*

The depth of construction and flexibility of the trusses in bending were such that an additional arching action was occurring between the two main piers, and contributing to the damage at the four half-joints. Localised rotation of each joint when heavy vehicles passed over the structure was also a consideration. High coefficients of friction at the sliding bearings resulted in significant restraint to these actions. Although the bearings were moving correctly, the original sliding surfaces were phosphor bronze on steel which were measured as providing a coefficient of friction of 35 per cent. (Note that this met the requirements of published guidance at the time the structure was designed).

Cracks up to 100 mm long were found to have formed in the upper and lower steel box members at the half joints. Repairs to the viaduct were completed in 1996.

The structure should be checked to ensure that the key components for its intended articulation are all functioning correctly, and are properly maintained. Bearings, expansion joints, and any other mechanical devices on the structure should be regularly lubricated to ensure that movement can take place at load levels below the design loads. A lubrication regime as part of the regular maintenance operations may be an effective way of prolonging the operating life of a sliding joint or bearing. At the same time, the inspection records should be examined to ensure that the structure is not moving in any unusual manner, which would suggest that one or more component could be experiencing extra loading. An example of this is given in the papers by Matthews and Ogle (1996a and 1996b) on Boston Manor Flyover. Summary details are given in Box 4.3.

Expansion joints should be regularly cleaned, and checked for freedom, direction and range of movement. It should be recognised that the joint may only lock-up at extremes of temperature which are unlikely to coincide with the time when inspection takes place. The joint should be examined for signs of abrasion or damage. Maintenance operatives should be specifically briefed to check for these signs during regular cleaning operations.

Bridge bearings should be checked in a similar manner to expansion joints, recognising that in addition to translation movement, a bearing will require rotational capacity. Older types of bearing will include roller bearings, which should be checked for evidence of seizure, or cracking in the rollers. Bearings may appear to be seized but care is necessary before restraint is removed because the restraint may be indicative of a larger problem, for example, movements of abutments or supports.

Modern bearings rely upon PTFE or phosphor-bronze plus PTFE combination materials to provide sliding surfaces with low friction. Typically coefficients of friction for these materials are in the range of four to six per cent. Earlier bearings used materials which have higher coefficients of friction, and although sliding movements may be observed, the forces to cause the movement may be quite large.

Further guidance on bearings is given in the *Steel designers manual* (Owens and Knowles, 2003), and in the publication by Lee (1990).

Other restraints to structures may occur owing to unintended effects in bracing. These restraints will generate unwanted loading effects that can lead to deterioration and damage if not properly allowed for. An example is given in Figure 4.8 (a). In this case the bracing (provided solely for stability during construction) forms a truss across the width of the deck, and attracts loading under certain traffic conditions. This will generate loading in the top and bottom cross-members and connections.

Having been provided only for construction loads, the bracing may not be capable of carrying the larger loads generated in service. Moreover, if the loading condition is regularly repeated, the connections may be subject to fatigue damage.

Better solutions are shown in Figures 4.8 (b) and 4.8 (c), where the arrangement is more flexible, and truss action across the width of the deck is not possible. Structures should be reviewed for such occurrences, and details identified for checking during principal inspections.

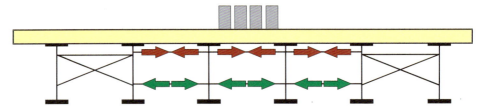

(a) Bracing across the width acts as a truss and attracts load.

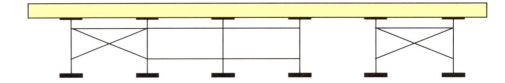

(b) Two pairs plus flange bracing for centre two.

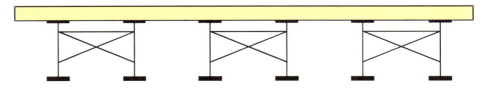

(c) Bracing of pairs of girders for erection.

**Figure 4.8**        *Configurations of transverse bracing*

## 4.4        GOOD PRACTICE

In this section good practice in preventative maintenance is briefly summarised. Additional and more detailed guidance is given by the UK Bridges Board (2005), the published papers from the bridge management conferences at the University of Surrey, Harding *et al* (1993, 1996 and 2000), and published papers from International Association of Bridge Maintenance and Safety (IABMAS) conferences <http://www.iabmas.org/>.

There are risks associated with any maintenance intervention, and there are considerable advantages in the use of a risk management approach in planning the work. The work should be properly defined, and the risks identified. Risks should be removed or mitigated by investigation. Residual risks should be shared between the owner (client) and designer rather than simply transferred. An appropriate level of investigation should be undertaken and this may include analysis, testing, observation and monitoring to calibrate the analysis. The option to renew the structure may be cheaper, and may have to be considered.

When it is decided to carry out a maintenance intervention, it is important to recognise that on older structures, there are likely to be hidden problems that are difficult to identify beforehand and only come to light as the work proceeds.

It is important to understand the bridge behaviour at both ultimate and serviceability limit states. Initial simplified analysis will frequently lead to identification of critical areas. More detailed analysis can follow if necessary. All analyses should be viewed critically, without having implicit faith in the results. The engineer should check that the analysis supports the observed behaviour, and if not, be diligent in understanding the reasons for the difference.

Design for remedial works should use solutions that are simple and flexible. The risk management process should lead the engineer to consider in some depth the constraints imposed by the remedial process, and allowance for those areas of the structure where condition cannot be confirmed until work commences.

In undertaking a review of the proposed scheme, the fitness for purpose should be considered not only in terms of the structural considerations, but also in terms of the legal constraints. It should be questioned whether the form of contract is likely to constrain the physical result and whether enough time has been allowed to complete the work safely and to a good standard. Are the penalties for late completion such that if any of the identified risks materialise during the job, it will be rushed and badly finished?

There is a considerable volume of knowledge and experience within the industry, and the following actions should be considered as part of the planning process:

- research similar solutions in the literature

- contribute to a collegiate approach and best practice web forums

- consider that dialogue and discussion may be preferable to formal reports

- question established practice in relation to the particular application

- keep records of information obtained for future reference by others.

# 5        Defects and deterioration

> This chapter deals with the different types of defects that can occur in iron and steel bridges:
> - material and manufacturing defects
> - fabrication defects
> - mechanical damage
> - corrosion
> - fatigue
> - environmental damage
> - damage from other external causes.

## 5.1      INTRODUCTION

This chapter provides a summary of the defects and processes of deterioration commonly encountered in iron and steel bridges, and their effects on the serviceability or ultimate strength. The defects are dealt with generically since many are, to differing extents, common to cast iron, wrought iron and steel.

Deterioration is defined as a reduction in performance over time. The performance measure is arbitrary and could relate to:

- a whole structure eg overall serviceability or ultimate strength

- a component eg the remnant strength of a riveted connection.

Deterioration is usually initiated at flaws in the materials or their protective systems which are subject to environmental and mechanical effects. In practice deterioration eventually results in defects which are big enough to be observed visually by a bridge inspector, for example failure of a protective coating which permits corrosion to develop. In this sense, deterioration is not the issue, but it is the defects which can be observed, recorded and repaired as part of the inspection and maintenance cycle.

In considering defects it is important to be able to identify the material that is being considered ie whether it is cast iron, wrought iron or steel. This may or may not be obvious and in an old bridge there may be a mixture of materials, which can bring its own problems. Identifying the material is important to successful treatment of the causes. Laboratory tests and field observations to determine material types are discussed in Chapter 6. In many cases, a metallic structure will have a protective coating, usually paint, which may need to be disturbed in order to aid the identification, and it is necessary to ensure that it is properly reinstated to avoid the development of local corrosion.

Defects can be inherent due to manufacturing or material anomalies. They can be isolated occurrences due to accident or vandalism, or can slowly increase in severity over time, depending on the environment.

It is reasonable to assume that all iron and steel bridges, particularly the older ones, have defects in one form or another. However only some of them will lead to deterioration and loss of performance. The defects which cause problems will also depend on the loadings being applied and the local environment of the structure. Not all visible defects or deterioration are significant from either an ultimate strength or serviceability perspective.

For example corrosion of the flanges (sometimes called pans) of steel toughing close to the supports does not necessarily affect its ability to carry load safely.

Defects or deterioration may not be visible but they can have a significant effect on the load carrying capacity. Hidden details are one of the biggest problems for the bridge inspector and assessor. This aspect is covered in more detail in Chapter 6, and is exemplified by the collapses of Inverythan Rail Bridge (Berridge 1969) and Kinzua Viaduct, see Appendix A2.3.

## 5.2     MATERIAL DEFECTS

This section is concerned with defects caused by the composition and metallurgical structure of the material or those produced during the production process. Such defects were prevalent in early versions of iron and steel but are unlikely to be present in the cleaner and more tightly controlled production processes now employed.

### 5.2.1     Brittle behaviour

Brittle behaviour, or the absence of ductility, is an inherent property of grey cast iron albeit there are now relatively few wholly cast iron bridges in service. In metals, plastic deformation occurs by dislocation motion. In body-centred cubic metals, such as low carbon steels, the flow stress required to move dislocations is temperature-dependent so that these materials exhibit a marked ductile to brittle transition most commonly at temperatures below ambient. Modern steels are usually available in a range of toughness and are selected depending on the application. For structures where the steel grade is unknown, the Charpy test may be used to determine toughness (see Section 6.2 for a description of the Charpy test).

Embrittlement of wrought iron with time was once considered a potential problem but there is no evidence to support this theory. There are sometimes problems with rivet heads cracking, which are likely to be a result of overheating the rivets followed by the working process of forming the head.

Hydrogen embrittlement can seriously reduce the ductility and load-bearing capacity of some metallic structures. The most vulnerable are high-strength steels. It involves the diffusion of hydrogen into the metal and can cause cracking and eventual brittle failures at stresses below yield. Examples where hydrogen embrittlement may be encountered include:

- welds
- hardened steels
- high-strength friction grip bolts.

### 5.2.2     Casting defects

Various casting defects were liable to occur in early cast iron, for example:

- blow holes due to poor venting of the mould
- residual stresses caused by differential cooling rates
- coarser and weaker material at the centre of the section
- contamination by sand becoming detached from the mould
- cold joints due to interruptions in casting

- cold-spots where earlier splashes of molten iron have cooled and solidified without subsequent absorption

- large variations in section thickness

- deformation and surface defects caused by damaged moulds.

### 5.2.3      Inclusions

The manufacturing process for wrought iron gave rise to the inclusion of slag (the impurities and residue from the furnace). These inclusions are worked through the metal during its manufacture and result in layered and fibrous strands throughout the metal which tend to align in the direction of working to form stringers, see Figure 2.2. This lamella structure significantly affects the properties of the metal in the direction of the grain and across it, rather like timber.

The fibrous inclusions are responsible for the characteristic way in which wrought iron can corrode, the so called puff pastry appearance.

The lower strength of the boundaries between the laminations of grains and inclusions can cause problems with welding and is one of the reasons why fillet welding to wrought iron is difficult but achievable.

Within castings, gas bubbles (draw holes), spatters of solidified iron and other impurities may have been trapped in the body of the metal. These will act as stress raisers and sites for initiation of cracking, and are often found to be the reason for cracked cast iron members. On Inverythan Rail Bridge a draw hole in a component of one of the main beams was hidden within a bolted connection so that it could not be seen by inspectors. This led to the bridge collapsing 25 years after construction. It was subsequently estimated that the draw hole was 250cc in size and caused the bottom flange of the beam to reduce in section by 25 per cent (Berridge, 1969). In this instance the loss in load-bearing section was the cause of collapse whereas there have been other cases where relatively small defects have had sufficient notch acuity to lead to brittle fracture, or the initiation and propagation of fatigue cracks.

## 5.3      FABRICATION DEFECTS

This section is concerned with defects introduced during fabrication of the as received material ie rolling, riveting and welding.

### 5.3.1      Incorrectly repaired castings

Cast iron components having defects such as blow holes were sometimes repaired in the foundry by being infilled with non-structural material sometimes referred to as scoria to give the appearance of being sound. A well publicised example is the use of Beaumonts Egg for castings in the first Tay Rail Bridge. It is unlikely that such critical defects have survived to the present day, but in practice they would be very difficult to detect on-site, particularly under a protective coating of paint. However, they can be removed under the action of grit-blasting or needle gunning when components are being prepared for painting.

An example of a defect in a bridge beam which was over 100 years old and had not been discovered until an inspection carried out in 2006, is shown in Figure 5.1. This defect penetrated the full thickness of the cast iron web so that when removed, daylight could be seen.

Note the corrosion staining to the ironwork

**Figure 5.1**    *A large defect filled with non-metallic material*

## 5.3.2    Surface defects

Surface defects can occur through use of chipped rolls which permit small and sharp upstands to be rolled onto the surfaces of a component. The deliberate rolling of details onto the component, such as the name of the manufacturer, introduce local stress concentrations that can, under some circumstances, act as initiation points for fatigue. Fortunately such details usually have relatively large root radii which have low stress concentrations, and failures from this cause are very rare.

## 5.3.3    Rolling

Millscale could be incorporated during the rolling process in wrought iron and early steelwork. The scale is usually left on the surface rather than becoming fully embedded. It can cause poor corrosion resistance because of the discontinuity in the surface and may result in pitting corrosion on the perimeter of the flake.

Cold rolling below the softening temperature of a metal can give rise to strain hardening and reduced ductility. In existing structures this would be difficult to detect although it may be the cause of other defects, such as cracking which are detectable.

## 5.3.4    Out-of-true rolling

In early iron and steel bridges constructed before design standards had been established, out-of-true rolling could occur. Where it has survived it is likely to be most noticeable and significant on secondary bracing members having a small cross-section relative to their length. The buckling resistance of compression members can be sensitive to any deviations from linear. Tension bars that are not straight will exhibit non-linear behaviour which could cause unwanted load distributions and require to be considered in the assessment of strength.

Rippling in plates subsequently used in webs may affect the resistance of the metal to in-plane shear stresses. Such deformation may need to be surveyed to allow for this defect in the strength assessment. In the time since standards have stipulated manufacturing limits, out-of-true rolling has no longer been prevalent.

### 5.3.5 Pre-cambering

Steel beams have sometimes been pre-cambered by cold-bending to compensate for subsequent deflections caused by self-weight and superimposed dead loads. Pre-cambering is not normally thought of as a defect but can result in strain hardening and loss of ductility.

### 5.3.6 Misalignment of rivet holes

Where plates are joined by rivets or bolts without having been match-drilled, then there is likely be some misalignment of the holes. However, provided the shank of the rivet or bolt could be drifted into place, then the mismatch would have no structural consequence.

Misaligned riveting is of more practical concern to the bridge engineer on occasions when it is necessary to remove rivets as part of a strengthening scheme. The effect of misalignment of holes is to make the process of rivet removal much more difficult and time-consuming and may not be apparent before work commences.

### 5.3.7 Welding

Many of the early examples of welded steel assemblages such as stiffened beams, exhibited minor buckling in panels between stiffeners due to residual welding stresses. These imperfections which can be seen in reflected light as ripples, can reduce the compressive stability. Design rules imposed after the recommendations of the Merrison Inquiry (Merrison, 1974), subsequently placed restrictions on these out-of-plane deflections.

In steel structures, problems can result from incorrect weld procedures during fabrication. The heating and cooling process can result in reduced ductility and defects such as sharp notches at the toe or root of the weld. Similar problems of reduced ductility and defects can also result from repairs carried out on-site by welding where damaged material has been worked after heating.

## 5.4 MECHANICAL DAMAGE

Mechanical damage may be due to the accidental, intentional or malicious damage of the structure. In many cases it can be a function of some weakness in the material in combination with a particular set of stresses to which the member has been exposed at some time in its history, for example constrained expansion during hot weather.

### 5.4.1 Impact damage

The accidental impact of vehicles into metal bridges (bridge bashing) usually results in damage to the bridge members by distortion and sometimes tearing or cracking of the metal plates. Where flanges, for example, have been bent out-of-plane, it is normal to ignore their contribution to the residual strength of the bridge (see Figure 5.2). In this example the damaged paintwork was not renewed and corrosion developed and spread from the areas no longer protected.

**Figure 5.2**

*Flanges bent due to vehicle impact. The impact also caused some of the paintwork to be cracked and spalled leading to subsequent corrosion*

In addition to the local effects of distortion around the point of impact, the global effects of the impact on elements such as the bearings and secondary or adjacent members will need careful examination. Damage caused by impacts is sometimes left unrepaired, as in the example in Figure 5.3, which has remained in this condition for some 50 years or more without leading to any serious consequences.

**Figure 5.3**     *Impact damage to vertical stiffeners of a wrought iron half-through bridge*

Other examples of vehicle strikes and subsequent repair work are given in Chapter 9 and Appendices A2.2 and A2.5.

## 5.4.2     Cracking and fracture

While there are relatively few wholly cast iron bridges remaining in use by vehicular traffic (Coalport Bridge is a notable exception, see Appendix A2.1), cast iron footbridges and cast iron components such as columns, parapets, edge beams and architectural details are still fairly common. There are also bridges having the external appearance of cast iron but their internal beams have been replaced by steel or concrete. Where used as a decorative casing, the material around the fixing points is likely to be the point at which cracks will start. Where cast iron details trap water, corrosion may develop unseen in the main structure, or crack the casting through freezing of entrapped water. Figure 5.4 shows the example of a cast iron bridge pier which has cracked and been repaired with steel bands.

**Figure 5.4**

*Cracking of a cast iron bridge pier*

An example of cracking occurred in the former cast iron pilasters over the piers of Hungerford Rail Bridge, London, as shown in Figure 5.5. These panels were not intended to be structurally significant but attracted stress and cracked. They were subsequently removed when the bridge was strengthened in the 1980s.

**Figure 5.5**    *Cracking in a pilaster above a pier of Hungerford Rail Bridge (photographed in 1979)*

If cracking is encountered in wrought iron or more modern welded steel structures, it is usually located at connections and likely to be a result of fatigue (see Section 5.7). However, cracking can also occur in modern welded steel when welds have been applied to high strength steel or large sections without adequate pre-heat. Uneven or rapid cooling of the weld can result in cracks which may propagate depending on the service stresses.

## 5.5        CORROSION

### 5.5.1      Appearance of corrosion

Corrosion is a common cause of deterioration and development of defects in iron and steel structures. Iron readily forms oxides in the presence of moisture and oxygen. The oxide form has a volume many times greater than the parent metal so that expansion occurs pushing the oxidised layer away from the parent allowing another layer to oxidise below. This expansion often causes corroded iron to look worse than it really is. General corrosion (as opposed to localised corrosion) usually occurs at a predictable rate. If the local environment remains unchanged and maintenance is carried out at appropriate times, the remaining life can be managed with some confidence.

In contrast to wrought iron and steel, cast iron has generally good resistance to corrosion evident in the performance of the cast iron in Pontcysyllte Aqueduct (see Appendix A2.7).

There are many good references eg by the National Physical Laboratory (2004), which provide detail on the chemistry of corrosion, and on the means of preventing it.

In wrought iron plates, the presence of slag inclusions (or stringers) coupled with the expansive action of corrosion, tends to cause delamination. Potentially, much of the overall thickness of the original section can be consumed by the corrosion process, resulting in knife edge flanges and wafer thin or holed webs.

Delamination can also be observed in some rivet heads in old structures. The presence of inclusions in the rivet material results in a similar layering effect. Once started, the corrosion can reach down into the body of the rivet and eventually cause the head to disintegrate.

Pitting is highly localised corrosion with metal being lost over small discrete areas. It is usually as a result of surface defects, which have been formed during manufacture such as rolled-in millscale, and provides a site for the corrosion to initiate. It may also be a result of the local breakdown or damage of the surface protection coating. Pitting corrosion may occur quite rapidly under damp conditions.

Contrary to common belief, the form of corrosion that affects the iron and steel commonly used in construction of bridges does not usually lead to sharp pits having stress concentrations sufficient to cause brittle fracture or initiate fatigue cracking, Larsson (2006).

### 5.5.2      Bi-metallic (galvanic) corrosion

Bi-metallic corrosion (sometimes referred to as galvanic corrosion) can occur when dissimilar metals are in contact in the presence of an electrolyte. The conditions for bi-metallic corrosion to occur are as follows:

-   the presence of an electrolyte bridging the two metals (the commonest is water)
-   electrical contact between the two metals
-   a difference in potential between the metals to enable a significant galvanic current to flow
-   a sustained cathodic reaction on the more noble of the two metals.

The rate of corrosion is affected by, among other things, the area ratio of the anode and cathode, ambient temperature and conductivity of the electrolyte. The degree of corrosion depends on the particular metals concerned and PD 6484 (BSI, 1979) provides a full guide to possible effects. The guide to bi-metallic corrosion published by NPL provides further detail (National Physical Laboratory, 2004). Table 5.1 gives an overview of the common metals found in structures in order of decreasing nobility (increasing risk of bi-metallic corrosion). Note that stainless steel and aluminium are not immune from galvanic corrosion.

**Table 5.1**     *Nobility of metals relevant to iron and steel bridges*

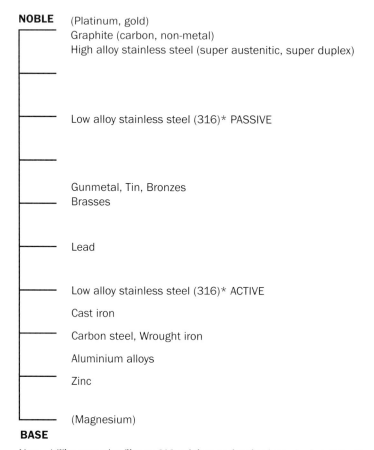

**NOBLE**     (Platinum, gold)
Graphite (carbon, non-metal)
High alloy stainless steel (super austenitic, super duplex)

Low alloy stainless steel (316)* PASSIVE

Gunmetal, Tin, Bronzes
Brasses

Lead

Low alloy stainless steel (316)* ACTIVE

Cast iron

Carbon steel, Wrought iron

Aluminium alloys

Zinc

(Magnesium)

**BASE**

Note     * The protective film on 316 stainless steel makes it passive but if the film is damaged the stainless steel becomes active.

Depending on the electrolyte, the cathode can form calcareous deposits which can act as insulation and decrease the rate of bimetallic corrosion. For example, steel corrosion can be decreased by an order of magnitude after less than a year's exposure.

Bi-metallic corrosion can be prevented by providing insulation between the dissimilar metals. This can be achieved by coating one or both of the metals with paint or, for example, by provision of non-conducting insulators such as polymer washers to separate a steel bolt from an aluminium alloy panel.

## 5.5.3     Stray current corrosion

It is possible for stray electrical currents to promote corrosion. This may occur for example when a direct current flows through an unintended path such as a fault in buried cabling. Particular consideration might be given to this as a possible cause of corrosion in railway underbridges carrying third-rail electrification.

## 5.6    DETAILS PRONE TO CORROSION

Some of the design details where corrosion has been found to be a problem in practice, and which should be investigated when carrying out an inspection, are given in the following sections.

### 5.6.1    Bearings and expansion joints

In many of the early iron structures there was no allowance for articulation and the designs, like masonry arches, relied on the structure being sufficiently flexible to be able to cope with thermal movements. The flexibility was usually at the connections between the superstructure and embankment, often causing eventual damage due to the cyclic movements. In later designs having bearings and expansion joints, regular maintenance is required to avoid the development of problems such as corrosion which can cause seizure and lead to the development of secondary actions and cracking (see also Section 4.3.8).

Bearings are invariably positioned in locations where detritus can collect, and corrosion can develop and hinder or prevent free articulation. Since bearings are designed to carry high loads, high stresses are generated and any inherent material defects can result in premature failure. Often bearings are not easily accessed for inspection, so failure may go undetected until a secondary and more visible effect is apparent. An example of a corroded bearing is illustrated in Figure 5.6.

Expansion joints can become seized-up by the action of detritus and corrosion as exemplified by the damage which occurred on Armstrong Bridge, Newcastle, described in Appendix A2.8.

**Figure 5.6**        *A corroded cast iron bearing*

## 5.6.2     Bearing pins

Where pinned connections have the flexibility to rotate (such as on hangers), localised corrosion can occur. This corrosion may not be visible but will result in increased stress in the pin. If the corrosion products then fuse the pin and the plates together, the pin can be subject to torsion in addition to the originally intended shear loads. Needless to say, such a combination of stresses can lead to premature failure.

Fretting is caused by the repeated relative movement of two components in forced contact. In effect one surface rubs away on the other. As well as abrasion, which is unlikely to be particularly significant, fretting continually opens up protective layers to the atmosphere, promoting corrosion. Situations where this might be anticipated are in the pin joints of suspension bridge hangers (when in working order) and elements supporting machinery in moveable bridges which could be subject to continuous high frequency vibration.

## 5.6.3     Riveted connections

In riveted connections that are insufficiently tight or receive inadequate maintenance painting to exclude moisture, corrosion can develop at the interfaces as shown in Figure 5.7. In extreme cases the expansion of the interfacial rust products can force the plates apart and fracture the rivets.

**Figure 5.7**     *Interfacial corrosion of a riveted connection*

Cracking around rivets or bolts may be a consequence of holes punched during fabrication. The action of punching leads to work hardening around the edge of the hole and the consequent reduction in toughness, coupled with the unfavourable geometry of the punched profile, increases the likelihood of fatigue cracking.

In current bridge specifications it is now required to carry out reaming of punched rivet or bolt holes, except under certain conditions, BS 5400-6 (BSI, 1999).

In cases when a rivet head is lost it should be replaced, ideally like-for-like but more conveniently, by a bolt.

## 5.6.4     Cast or built-in connections

Where members or parts are cast-in to concrete or built-in to masonry, corrosion can be severe at the interface and some distance into the containing material. A similar situation occurs where masonry or concrete is placed against metal parts. This is a hidden defect that can be overlooked and in any case is difficult to inspect.

An example of corrosion of a cast-in connection is given for base connections of the vertical columns supporting a railway footbridge in Figure 5.8 (a) and (b). The connections had previously been repaired by casting concrete blocks at the bases but corrosion developed to the extent that daylight could be seen through the corroded areas at the feet of the columns close to the concrete. This was despite the upper surface of the concrete being sloped away from the steelwork to prevent rainwater collecting at the interface. A second repair was made by removing the concrete and strengthening the steelwork with bolted cover plates. The concrete was then replaced.

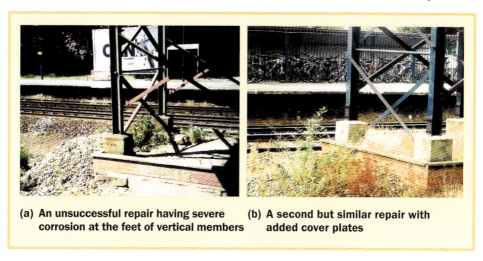

(a) An unsuccessful repair having severe corrosion at the feet of vertical members

(b) A second but similar repair with added cover plates

Figure 5.8      *Repairs to columns supporting a footbridge*

An alternative connection that has commonly been used has a flanged detail at the base of the column which is bolted to the concrete. However, this transfers the susceptible interface to the bolts which would require removal for inspection and may in the long-term fail without warning.

### 5.6.5      Tie-rods

Tie-rods, particularly the transverse tie-rods in jack arches, commonly become corroded through leakage from the deck above and in extreme cases become severed so that their structural action is lost.

An example of a corroded tie-rod is shown in Figure 5.9.

Figure 5.9      *Corroded tie-rod*

## 5.6.6    Poor design and maintenance

Metallic structures are usually protected from corrosion by a coating such as paint or metal spray, see Chapter 11. However the integrity and performance of the coating will be greatly influenced by the original design and the subsequent maintenance. Where the design makes it difficult for maintenance the problem is likely to be even greater. Design details where detritus can collect are likely places where corrosion can occur.

Examples of poor detailing include:

- locations where the pavement or road surfacing butts up to girder webs, Figure 5.3

- exposed bottom flanges on filler beam decks and jack arch decks, Figure 5.9

- welded additions to existing steelwork, Figure 5.10 (a)

- cross-connections, welded, bolted or riveted, Figure 5.10 (b)

- lattice-work to flanges, Figure 5.10 (c).

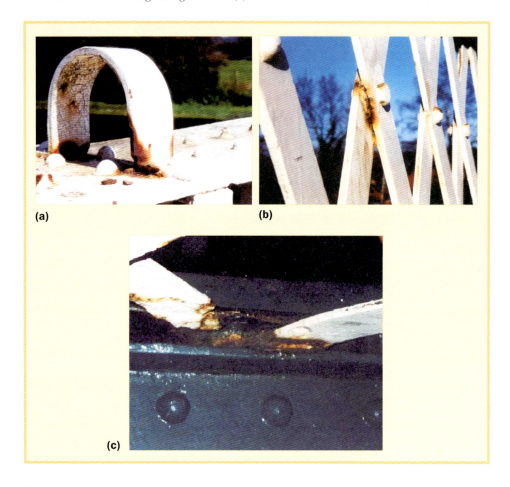

(a)

(b)

(c)

Figure 5.10         *Examples of corrosion caused by susceptible detailing*

It is well known that poor maintenance of dirt traps (at re-entrant details) is a common cause of corrosion. The build-up of debris and detritus on iron or steel members reduces the life of the coating through various mechanisms, but the retention of moisture in contact with the surface is the main factor. Designs that are self-cleaning (inclined surfaces that are washed by rain for example) and easily maintained (provided with built-in access for cleaning) are important factors in extending the life of the structure.

Water management on the bridge arguably has the biggest influence on its long-term durability. Where water is allowed to pond or percolate through the structure, moisture is retained providing ideal conditions for corrosion to start, particularly in corrosion traps. Positive drainage, with adequate and accessible points for cleaning, and a sound waterproofing system should be provided when opportunities arise during refurbishment, repair or strengthening work.

## 5.7 FATIGUE

Fatigue is the phenomenon in which a metal component fails as a result of the repeated application of loading cycles. The stress ranges induced by the loading cycles can be well below the elastic limit of the material and yet failure can eventually occur. Failure occurs as a result of the growth of a crack which eventually reaches a critical size, at which point failure either occurs as a result of a net section stress effect or in a brittle manner due to the stress concentration associated with the crack tip. A fatigue crack has two distinct stages:

1   Initiation.

2   Propagation.

Fatigue cracks initiate at the surface of a component (assuming of course that the component does not contain any significant subsurface defects) and the quality of the surface finish plays a primary role in sensitivity to fatigue. In the extreme, small scratches or tooling marks, invisible to the naked eye, can initiate fatigue cracks in aerospace structures but the types of bridge structure designed to date are considerably less susceptible.

When there is doubt about whether a crack has been caused by fatigue, and the crack surface is accessible, the mode of failure may be identified from its appearance. Under optical magnification, the presence of striations caused by the crack propagation process are evidence of fatigue. However, the services of a fatigue expert should be obtained as fatigue can also occur without producing discernible striations.

### 5.7.1 Details prone to fatigue

In iron and steel bridges fatigue invariably initiates at poorly designed details. Examples of fatigue prone details designed in the past include:

- lightweight bracing subjected to vibrations in wind or traffic
- short hangers at mid-span of suspension footbridges and road bridges
- welds, typically connections to tension flanges and in lightweight orthotropic steel decks
- rivet and bolt holes.

The classification of details according to their fatigue life is given in BS 5400-10 (BSI, 1980) and BS EN1993-1-9 (BSI, 2005).

Guidance on the detection and monitoring of fatigue cracks is given in Chapter 6.

### 5.7.2 Incorrect modification

Modifications carried out in the past in order to repair or strengthen a structure, or to make way for services or other features, can have deleterious consequences if not properly considered and supervised. Classic examples are plates welded onto flanges which can then induce cracks that will grow under fatigue loads. Also, holes cut in webs to facilitate services. Any cutting of steel is likely to leave imperfect edges from which cracks may propagate given (un)favourable conditions of stress.

## 5.8 EXTERNAL CAUSES OF DAMAGE

### 5.8.1 Environment

Aggressive environments such as marine or coastal regions are well known for promoting corrosion in iron and steel bridges. Other examples are on bridges above railways in the areas exposed to the exhaust blast from locomotives and in the vicinity of heavy industrial plants and power stations. The corrosive action of de-icing salts on highway structures is the most common source of problems.

### 5.8.2 Timber

Timber members or decking naturally produce acetic acid in varying concentrations depending on the species of wood and length of time when exposed to damp conditions, see BRE Digest 301 (BRE, 1985). This acid can affect protective coatings causing breakdown and initiating corrosion, see Figure 5.11. Unprotected metals can be attacked directly by the weak acid and also through electro-chemical corrosion.

**Figure 5.11**     *Corrosion of steel crossbeams aggravated by timber decking*

### 5.8.3 Wildlife

Birds, particularly pigeons, like to roost on the underside of bridge decks. Their droppings can build up on bottom flanges of steel beams and cause damage due to the corrosive action of the guano, an example is shown in Figure 5.12. The build-up also forms poultices that trap water and accentuate the corrosive action. Defensive actions such as using netting to prevent access are sometimes thwarted by nature lovers cutting the nets, possibly to free pigeons that have become trapped, as shown in Figure 5.13.

**Figure 5.12**    *Build-up of pigeon droppings*

**Figure 5.13**    *Bird netting damaged, possibly by well meaning nature lovers*

Measures to prevent damage from pigeons are described in Section 4.3.6

## 5.8.4    Fire

Bridges are sometimes damaged by fire due to crashed vehicles or vandalism. If repairs are not carried out comprehensively it is possible that iron or steelwork may have been left in a weakened state.

Strength loss for steel is generally accepted to begin at temperatures of about 300°C and become more significant above 400°C. At 550°C steel retains about 60 per cent of its room temperature yield strength. This is usually considered to be the failure temperature for structural steel. In practice this is a conservative assumption as the insulating effects of concrete slabs and the restraining effects of connections cause failure temperatures to be higher.

When exposed to high temperature (over 600°C), steel may suffer some deterioration in residual properties on cooling. Temporary heating to a member eg from a car fire or an act of vandalism, should be investigated to determine the need for any remedial work or replacement to maintain the integrity of the structure.

The temperature characteristics of wrought iron are similar to steel. Strength loss of cast iron occurs above 200°C.

## 5.8.5 Vandalism

Iron and steel bridges are subject to a variety of forms of vandalism:

- parapets, particularly aluminium alloy, are liable to be removed. Replacement of cast or wrought iron parapets can be expensive

- steel doors into box girders or hollow piers are sometimes forced open or torn off and internal damage is caused

- graffiti is common and can be particularly difficult to remove from weathering steel

- combustible material stored or fly-tipped beneath bridges are liable to be set on fire.

# 6 Inspection, testing and monitoring

> This chapter deals with methods of inspection and monitoring iron and steel bridges. The various laboratory and *in situ* tests are described.

## 6.1 INSPECTION

Bridge owners have a duty of care to ensure that their bridge stock is maintained in a safe and serviceable condition. Serviceability is typically defined as "the ability of structures to fulfil, without restriction, all the needs which they are designed to satisfy", BA 57/01 (TSO, 2001). This duty of care is the driver behind the implementation of any inspection regime for a bridge.

Inspections may be considered as falling into two classes:

1   Inspections for condition.

2   Inspections for strength assessment.

although both seek to ensure the serviceability of the structure.

Inspections for condition are usually limited to the visual observation and monitoring of defects. This is sometimes supported by on-site testing, although in the case of iron and steel bridges the number of simple on-site tests is limited (see Section 6.2). Observations and test results are recorded in a standardised fashion on inspection pro-formas. However, the reporting systems adopted by different owners vary, particularly in the way that the condition of bridge elements is described.

Inspections should provide data on bridge geometry and dimensions, and evidence of any visible deterioration. Metallic bridges are generally robust and typically show evidence of distress in advance of structural failure, provided it is known where to look. The routine process of periodic visual inspection, and more detailed inspections on a less frequent basis, supported by additional investigations as required, is a potentially adequate strategy for ensuring fitness for continued service. However, deterioration and loss of performance of metallic bridges is not always obvious and it is important to consider hidden variations in quality and construction when considering bridge characteristics for structural assessment.

There are assumptions implicit in taking a bridge at face value, and where the risks associated with such assumptions are not tolerable it is necessary to carry out further investigative work to achieve the necessary confidence in visually-based inspection data. Such investigations can provide additional information on specific aspects of the bridge's structure and its behaviour, hidden details, materials type and condition and deterioration, using a variety of intrusive, non-destructive and analytical techniques. Investigation techniques are discussed further in Section 6.2.

## 6.1.1      Inspection regime

Bridge owners regularly review their standards and codes of practice for setting out their various inspection regimes. As a result, this whole area is in a continuous state of flux. What follows is a snap-shot of the position with respect to the bridge owners listed, at the time of writing.

Inspections should be sufficiently regular such that the possibility of any particular bridge becoming unsafe or suffering an unacceptable loss of serviceability between inspections is acceptably low. The regularity of inspection should also be sufficient to ensure that any deterioration is detected early to permit the most cost-effective remedial action. All the large UK bridge-owning authorities have their own internal policies on the regularity of inspections, including:

- for Network Rail (NR) bridges, examination types, requirements and intervals are set out in Specification NR/SP/CIV/017 (Network Rail, 2004b)

- for London Underground (LU) bridges, inspection types, requirements and intervals are set out in Engineering Standard 2-01304-002 (LUL, 2006)

- the requirements of the Highways Agency for inspections and intervals are set out in BD 63/07 (TSO, 2007)

- British Waterways (BW) carry out inspections generally in accordance with the principles of BD 63/07 but the inspection frequency is variable and determined from a risk based matrix

- local and unitary authorities invariably use HA documents.

The terminology and frequencies of inspection vary between the main UK infrastructure owners, although a similar basic principle of four types of inspection, varying in terms of typical objectives and methodology, is a common approach as set out in Table 6.1.

Fixed-schedule inspection and assessment schemes have some negative consequences, since valuable resources may be spent on bridges that are known to be in excellent condition, whereas bridges in poor condition may not be inspected as regularly as necessary. A measure of flexibility is desirable, based on a proper assessment of risk, so that resources can be directed where they will be most effective, while ensuring the prime objectives of safety and functionality. Subject to the policy of the bridge owner, limited variations in inspection frequencies may be permissible depending on the use, type, condition, deterioration and accessibility of the bridge, and the perceived effectiveness of the inspection itself. This requires justification, typically through a risk assessment process to demonstrate the acceptability of the proposed inspection frequency. This approach is considered advantageous, but the risks associated with reduced inspection frequency need to be adequately assessed on a structure-by-structure basis.

An increase in the intervals between inspections may be considered acceptable if it has been demonstrated that:

- the condition of the structure is good and there is no potential for rapid deterioration

- the capacity of the structure exceeds the applied loading by a significant margin

- there is a high level of confidence in the results of inspections and assessments

- it is not envisaged that there will be any significant changes in use, loadings or environment which might affect the bridge detrimentally

- the potential modes of failure of the bridge are understood and there is adequate confidence that the proposed inspection type and frequency can adequately identify structural distress in advance of failure, or that the consequences of failure are low

- the likelihood of incidents which might affect the capacity of the bridge structure is low (eg bridge strikes, excessive traffic or environmental loading).

Conversely, a decrease in the intervals between inspections may be necessary if:

- the condition of the structure is poor or there is the potential for rapid deterioration

- the capacity of the structure barely exceeds the applied loading or if there are restrictions on its use

- the level of confidence in the results of inspections and assessments is not high

- changes in the use, loading or environment of the bridge, which might detrimentally affect its performance, are envisaged

- the potential mode of failure of the bridge is poorly understood and there is inadequate confidence that the current inspection regime can identify structural distress in advance of failure

- the consequences of failure are perceived to be particularly high

- the likelihood of incidents that might affect the capacity of the bridge structure is high.

Where a risk assessment is used to justify an increase in intervals between inspections, it is particularly important that it is updated with current data, reviewed and re-assessed at suitable times.

## 6.1.2 Planning

Proper planning of an inspection is important. Often the time on-site will be limited, short duration track possessions for the inspection of railway bridges being an obvious example. It is essential that the time which is available for the inspection is efficiently used.

All available drawings of the bridge should be studied so that members of the inspection team are intimately familiar with the form of construction and details of the bridge prior to visiting site. Ideally these should be as-built record drawings rather than design drawings. If there are no drawings, as is sometimes the case with older structures, then it is essential that a preliminary visit to the bridge be made prior to the inspection proper. Such reconnaissance visits are also useful for identifying whether vegetation or other obstructions need to be removed before the inspection, and for determining the best method of access to the deck soffit and other parts of the bridge.

If the land below the bridge is private then the reconnaissance visit may provide an opportunity to approach the land owner for permission to enter and to establish the means of access.

The inspection team should study any available maintenance records and, as a minimum, the previous inspection report. Where there is evidence of ongoing deterioration or other long-term problems with the bridge it may be useful to study earlier reports as well to assess whether the rate of deterioration is increasing.

The inspection team should be familiar with the utilities carried by the bridge. It is not always necessary to know their nature but as a minimum the number and locations of all ducts should be known so it may be quickly seen if anything new has been added since the last information was recorded. If the intention is to dig into the structure or undertake other intrusive investigations then it is important to know the nature of the utilities (gas, electricity, water, telecom) and their depth, to avoid either harm to those doing the digging or damage to the services. If not already done, their positions should be recorded to aid any future intrusive work on the bridge.

Wherever possible it is recommended that inspection pro forma are prepared beforehand. These should provide space to record all of the structural elements as well as the connections between them. They should also contain simple line drawings of the bridge in plan and elevation to enable the locations and orientations of all photographs to be recorded. Without this simple device it is all too easy to return to the office with numerous detailed photographs of corroded steelwork and other interesting details, and in the cold light of day realise that it is not clear which part of the bridge the photographs relate to.

Lastly the manager planning the inspection should make sure there are sufficient experienced staff available to undertake the inspection (see Section 6.1.3). The manager should be an experienced bridge examiner or engineer and be able to adjust the composition of the inspection team depending on what is found.

Once all of the above information and preparation is in place, a method statement that summarises all the relevant information should be prepared, and agreed by all parties. The level of detail given should be appropriate to the complexity, circumstances and type of inspection. Normally, the following minimum level of information should be included in any method statement:

- details and programme of the work to be undertaken
- equipment required
- methods of access to be used
- traffic management details
- the risk assessment including safe procedures for dealing with hazards
- health and safety requirements including equipment for personal protection
- the resources and competence of the staff to be employed
- contact details for key individuals involved in the inspection
- planned working times
- temporary works to be employed
- protection from highway, rail, waterway and other traffic
- requirements for action by others
- any co-ordination or notification required
- any environmental impacts of the work
- emergency arrangements in the event of an incident
- emergency arrangements in the event that the structure is found to be unsafe.

**Table 6.1** *Inspection requirements of principal UK bridge owners*

| Inspection level | Title | Scope | Interval |
|---|---|---|---|
| Routine cursory inspection | Superficial inspection (HA)<br><br>Length inspection (BW) | Cursory visual check for deficiencies which might lead to accidents or increased maintenance. Part of the day-to-day surveillance by bridge owner's staff (not necessarily trained inspectors) in the course of their normal duties. | When staff visit the bridge site<br><br>Monthly (BW) |
| Routine visual inspection | General inspection (HA)<br><br>Visual inspection (NR)<br><br>General inspection (LU)<br><br>Annual inspection (BW) | General inspections are defined in the *Management of highway structures – a code of practice* as comprising "a visual inspection of all parts of the structure and, where relevant to the behaviour and stability of the structure, adjacent earthworks or waterways that can be inspected without the need for special access or traffic management arrangements" (UK Bridges Board, 2005). | Two years after last (HA)<br><br>One year general or principal inspection (NR, LU)<br><br>Annual inspection by the length inspector and a certified annual inspector (BW) |
| Routine detailed inspection | Principal inspection (HA, LU, BW)<br><br>Detailed examination (NR) | A close examination, within touching distance, of all accessible parts of a structure utilising suitable access and/or traffic management works as necessary. Visually based but can be supported by measurement and simple testing (eg hammer-tapping) to gather additional data. | Normal intervals:<br><br>Four years (LU)<br><br>Six years (NR)[1]<br><br>Six years (HA)[2]<br><br>10 years max (BW) |
| Non-routine inspection | Special inspection (HA)<br><br>Additional examination (NR)<br><br>Defect advice inspection (LU)[3] | An inspection undertaken outside the normal regime either from a need to look in more detail at a particular part of the structure, or following an event such as a severe flood or vehicular impact.<br><br>A visual inspection can be augmented by specialist techniques (*in situ* testing, sampling for laboratory analysis, load testing) as required. | As required to investigate a particular feature or problem |

**Key**  Network Rail (NR), Highway Agency (HA), British Waterways (BW), London Underground (LU).

1    Where structural parts of the bridge are underwater, and where the depth of water prevents a visual examination, the normal interval between detailed inspections is three years.

2    Intervals can exceptionally be up to 10 years.

3    London Underground also require special inspections which are regular visual inspections carried out at short intervals for metallic structures awaiting repairs.

## 6.1.3    Competence of inspection personnel

It is vital that inspection personnel are equipped with the skills, knowledge, experience and supervision to adequately perform their duties, commensurate with the complexity of the task, and are supported with the necessary resources. It is also necessary that they have an adequate level of understanding to be able to judge when emergency measures are required for safety reasons.

Routine visual inspections are carried out by bridge inspectors or examiners, who may also be involved with the day-to-day maintenance of these bridges. Such staff perform the function of a trained pair of eyes able to spot obvious signs of damage and distress, and often have a good understanding of the requirements for routine maintenance and straightforward repairs.

The basic qualities of a good inspector are:

- a knowledge of safe working practices and access requirements for inspection

- experience of the techniques and tools available, and an understanding of their use and limitations

- an adequate understanding of the construction, materials and behaviour of metal bridges structures

- a knowledge of the causes of structural defects and deterioration of metallic materials

- an adequate understanding of the modes of failure of metal bridges, and the ability to recognise and interpret features which might require urgent action

- the ability to make and record objective observations accurately, clearly and consistently.

For a novice inspector to attain these qualities and become fully effective it is likely that he will require some formal training in addition to experience gained by working alongside a well practiced and trusted inspector to allow the transfer of knowledge and skills. In certain situations specialist training and skills may be required, for instance where inspections require roped access, working in confined spaces or underwater. Some of the major infrastructure owners specifically define the necessary standards of competence for bridge inspectors and provide training schemes which lead to formal qualifications in this respect to ensure compliance. For example, Network Rail set out the level of competence required in their Standard NR/SP/CIV/047 (Network Rail, 2005). London Underground Limited require inspectors to be technician members of the Institution of Civil Engineers.

## 6.2 TESTING

### 6.2.1 The value of testing

Testing is often required to determine the parameters needed for undertaking a structural assessment or when it is necessary to gather information on parts or features of the bridge that are not readily obtained by visually-based inspection techniques.

There are various techniques for obtaining material property data from old bridges. Very often the original material specification is unknown. Even if the original specification is known, the actual properties will usually be better than the minimum specified values. This applies particularly to yield and tensile strengths which are the most important properties for determining buckling and rupture strengths of a member.

In some cases, particularly where welding repair or strengthening is to be carried out, there is likely to be a need to check the material toughness, usually by Charpy tests. In this same context it is essential that the weldability of the material is properly evaluated, so that suitable procedures can be drawn up and tested (preferably on similar or sample material). The required welding data may include chemical composition, cleanliness, hardness etc.

In the absence of such data a conservative assumption will always have to be made. Frequently this may lead to expensive strengthening or even restrictions on the method of strengthening. Alternatively a lack of appreciation of the low level and range of quality often exhibited by the older irons and steels, when compared with their modern counterparts, can often lead to expensive mistakes when attempting repairs.

For these reasons, the benefit obtained from a few simple tests on the material in the structure itself is usually well in excess of the cost of such testing.

Table 6.2 provides a list of commonly used tests and the circumstances in which they provide useful data for the structural assessor. Most of the tests are suitable for wrought iron and steel, but less for cast iron. More than half of the tests can be carried out *in situ* with varying degrees of disturbance to the material. Nearly all can be conducted in the laboratory on samples extracted from the bridge.

The range of investigative techniques employed for these purposes is summarised in Table 6.2, and each technique is discussed in more detail in Sections 6.2.3 and 6.2.4.

**Table 6.2**   *Specialist investigation, testing and monitoring techniques for bridge investigation*

| Property | Type of test | Materials | | | Test location | |
|---|---|---|---|---|---|---|
| | | CI | WI | MS | *In situ* | Lab |
| Material identification | Chip test | ✓ | ✓ | ✓ | ✓ | × |
| | Spark test | ✓ | ✓ | ✓ | ✓ | × |
| | Surface etching | ✓ | ✓ | ✓ | ✓ | × |
| | Chemical analysis | ✓ | ✓ | ✓ | × | ✓ |
| Yield strength | Tensile test | × | ✓ | ✓ | × | ✓ |
| | Compression tests | × | ✓ | ✓ | × | ✓ |
| Hardness | Vickers, Brinell | × | ✓ | ✓ | ✓ | ✓ |
| Tensile strength | Tensile test | ✓ | ✓ | ✓ | × | ✓ |
| Ductility | Tensile test | × | ✓ | ✓ | × | ✓ |
| Compressive strength | Compression test | ✓ | ✓ | ✓ | × | ✓ |
| Notch toughness | Charpy test | × | ✓ | ✓ | × | ✓ |
| | Fracture toughness | × | × | ✓ | × | ✓ |
| Fatigue resistance | Crack growth | ✓ | ✓ | ✓ | × | ✓ |
| Weld quality | Macro examination | ✓ | ✓ | ✓ | ✓ | ✓ |
| | Micro examination | ✓ | ✓ | ✓ | × | ✓ |
| | Hardness | ✓ | ✓ | ✓ | ✓ | ✓ |
| Weldability | Chemical analysis | ✓ | ✓ | ✓ | ✓ | ✓ |
| | Hardness | ✓ | ✓ | ✓ | ✓ | ✓ |
| | STRA testing | × | ✓ | ✓ | × | ✓ |
| State of stress (dead load and residual | Centre hole | ✓ | ✓ | ✓ | ✓ | × |
| | Trepanning | ✓ | ✓ | ✓ | ✓ | × |
| | Sectioning | ✓ | ✓ | ✓ | ✓ | × |

**Key**   CI = cast iron, WI = wrought iron, MS = mild steel

✓ = applicable and useful
× = not applicable

Laboratory tests by definition require removal of material.

All of the investigations in Table 6.2 require a full consideration of programming and timing, health and safety environmental issues, planning, and preparation issues similar to those for inspections, as discussed in Section 6.1.2.

## 6.2.2     Material identification

In the majority of cases it will be possible to tell, by using a combination of visual inspection and some limited background knowledge, whether a particular part of a bridge is of steel, wrought or cast iron, as discussed in Chapter 2. Steel was introduced around 1880, and bridges from about 1880 onwards were likely to be steel although there were isolated wrought iron examples later. Cast iron was used for columns up to the early 1930s. Prior to about 1880 metallic bridges were exclusively iron. There is a period from about 1880 to c1914 where a particular component may possibly be of either iron or steel.

In beam sections there are usually sufficient clues to enable the inspector to differentiate with some certainty between wrought and cast iron and steel, whether these be in the form of fixings used, the proportions of the sections (unequal top and bottom flanges in cast iron beams due to different compressive and tensile strengths) or the degree of corrosion (cast iron corrodes much less readily than either wrought iron or steel, and wrought iron typically displays a characteristic fibrous or laminated structure when severely corroded). In addition cast iron can often be identified by mould marks, a variable section along the length, and holes left on its surface from the casting process.

When any doubt remains, it is necessary to test the metal to determine its type. There are several tests which can be performed *in situ* to aid in material determination:

- surface etching
- the spark test
- the chip test
- chemical analysis.

### Surface etching

Surface etching is a relatively straightforward procedure involving grinding, polishing and acid etching a part of the surface of the component and then examining the surface with a microscope (see Figure 6.1). From an examination of the microstructure of the material, the trained eye can readily identify whether it is cast or wrought iron or steel.

Cast iron is identified by the presence of dark graphite flakes in the etched surface. Wrought iron is identified by the presence of slag stringers. Steel is characterised by a granular structure mostly of pure iron with small quantities of pearlite. Figures 2.1 and 2.2 illustrate the characteristic appearance of etched sections.

**Figure 6.1**  *In situ identification of cast iron following grinding, polishing and acid etching*

## The spark test

The spark test involves holding a rotating grinding wheel against the surface of the component and noting the spark pattern produced. The chemical composition of the metal has a marked effect on the spark pattern. Cast iron produces an extremely short bushy spark stream with extensive branching. Wrought iron produces almost no branching but a spark trail typically twice the length of cast iron. Steels produce even longer trails, with low carbon steel producing the longest but with less branching than high carbon steel.

The two principal drawbacks of the spark test are:

1    It requires a high degree of experience to rely on it in isolation from other testing.

2    The use of handheld tools means that repeatability is not guaranteed as the size and shape of the spark trail depends on the speed of the wheel, the cleanliness of the wheel and the pressure on the test area.

## The chip test

The chip test, as the name implies, involves removing a small amount of the metal to be identified using a sharp cold-chisel. The colouration, fracture surface and the shape of the edges of the chip taken together can aid identification (see Table 6.3).

**Table 6.3**  *Material identification using the chip test*

| Metal | Fracture colour | Fracture appearance | Chip characteristics |
|---|---|---|---|
| Cast iron | Dark grey | Crystalline | Small, brittle, rough |
| Wrought iron | Bright grey | Fibrous, non-crystalline | Long, smooth |
| Steel | Light grey | Finely crystalline | Long, smooth |

Given the level of subjectivity and practical difficulties inherent in the spark test and subjectivity and destructive nature of the chip test, acid etching is considered the preferred method.

**Chemical analysis**

Chemical analysis of older materials is useful for identifying the type of material and, if it has been or is to be welded, what the weldability is likely to be.

The most common method is to use optical emission spectrometry (OES). This will provide the proportions of all the elements necessary to identify the material and its weldability. It is best done by extracting samples of material from the component and sending them to a laboratory for analysis. The samples can be obtained from drillings, chips or small slivers of material amounting to about 15 g weight. These will be re-melted before analysis.

Alternatively a sample off-cut of intact material with a flat surface at least 12 mm × 12 mm in area (and 2 mm thick) can be used directly.

It is usually best to sample from more than one location and preferably through the thickness. It should be noted that samples taken for other types of mechanical testing described above can be used for chemical analysis.

There are portable instruments for obtaining analysis *in situ*. The OES method can be used, but the equipment is expensive and the results are probably not as accurate as those carried out in the laboratory. There are cheaper site instruments based on X-ray fluorescence, but these will not give data for carbon and sulphur, (which are of particular interest).

In summary, the chip and spark tests can be used by experienced practitioners but even then can only make a high level distinction between material types. Surface etching and/or chemical analysis are both more reliable and more fundamental.

### 6.2.3 Non-destructive tests

**Hardness and strength testing**

Field hardness testing is a rapid method of evaluating the quality and variability of material in a structure. In addition, for steel bridges it may be used to estimate the tensile strength of the metal as there is a close relationship between hardness and strength in steels. This correlation is much less close for cast and wrought iron, and the test should not be used to estimate strength in these materials. However, BS 1452 (BSI, 2001) offers ranges of Brinell hardness for different grades of grey cast iron.

The Vickers hardness test involves pressing a pyramid diamond indenter into the test piece under a load of 10 kgf and measuring the size of the indentation. The size of the indentation is converted to a Vickers Hardness Number (HV/10) and from this to a tensile strength Fu by the approximate relationship:

$$Fu \ (N/mm^2) = 3.15 \times HV/10$$

However, the ultimate tensile strength given in the above relationship is rarely of use to the engineer and approximate relationships between yield strength and ultimate tensile strength have been proposed for steel viz:

$$Fy = 0.55 \times Fu$$

The Brinell hardness test is similar to Vickers but employs a hardened steel ball instead of a diamond pyramid.

## Flaw detection by magnetic particle testing

Magnetic particle testing is a non-destructive testing technique used for the detection of surface breaking or near surface breaking flaws. A typical use is for the detection of cracks in welds. It relies on local magnetisation of the component or surface being tested and so it can only be applied to ferromagnetic materials. If a defect is present on the surface or just below the surface and it is aligned in a suitable direction relative to the direction of the magnetic flux ie not parallel to the flux, then flux will leak from the sides of the defect.

Magnetic particles allowed to flow over the magnetised surface will be attracted to the flux leakage in the same way that iron filings are attracted to the flux from a bar magnet and the concentration of particles at the crack permits visual confirmation of the crack.

Although a magnetic powder can be used, the particles are often suspended in a liquid hydrocarbon to enhance their fluidity, and in many cases they are coloured to provide even greater contrast. For situations where the highest resolution is required, fluorescent coated particles are used in conjunction with an ultraviolet light source.

The success of this technique relies on the operator to induce sufficient magnetic flux in the surface being tested. The lighting conditions, contrast media and orientation of the defects relative to the induced flux are also important. Finally, the operator must know what to look for.

The success of this technique depends on the surface condition of the area being inspected, with defects smaller than about three times the surface roughness unlikely to be detected. Under optimal conditions, and with good quality surfaces, defects as small as about 0.5 mm can be detected with depths as little as about 0.02 mm.

An example of crack detection using this method is given in Appendix A2.

The relevant British Standard for magnetic particle testing is BS EN 1290 (BSI, 1998a).

## Flaw detection by dye penetrant testing

Dye penetrant testing is one of the earliest forms of non-destructive test, and in an early form it was used as long ago as the 1930s. It is used for detecting surface breaking flaws. It relies on capillary action drawing a liquid into the flaw, so it can only be used on a component or material which is non-porous.

To perform the test a layer of a penetrating coloured liquid is applied over the surface of the component. After a preset dwell time the surface is then carefully cleaned and dried. An absorbent developer (eg chalk powder suspension) is then applied over the clean dry surface. This draws out dye which has penetrated any flaws, and produces a stain on the surface. Fluorescent dye can be used in conjunction with ultraviolet illumination to make identification easier.

The success of the test relies on the operator to apply the test according to written procedures, the suitability of the procedures, and on the skill of the operator and lighting conditions when the surface is inspected.

The accuracy of dye penetrant testing depends upon the surface condition of the area being inspected, with defects smaller than about three times the surface roughness being unlikely to be detected. Under optimal conditions, and with good quality surfaces, defects as small as about 1.0 mm can be detected with depths as little as about 0.05 mm.

Dye penetrant testing can be used to identify the following:

- grinding cracks
- heat-affected zone cracks
- poor weld penetration
- weld cracks
- heat treatment cracks
- fatigue cracks
- hydrogen cracks
- inclusions
- laminations
- micro-shrinkage
- gas porosity
- hot tears
- cold joints
- stress corrosion cracks
- intergranular corrosion.

The relevant British Standard for dye penetrant testing is BS EN 571-1 (BSI, 1997).


## Flaw detection by eddy current testing

Eddy current testing is a non-destructive testing technique used for detecting surface breaking and near surface flaws in metallic components. It can also be used to a limited extent to measure changes in surface hardness, for the measurement of coating thicknesses and for heat damage identification by measuring changes in conductivity.

The basic principle of eddy current testing is that a probe is placed near the surface of the component to be tested. This probe contains an electrical coil which generates a magnetic field around the probe. By the process of electromagnetic induction this magnetic field induces a localised electric current in the component. This induced current produces its own magnetic field which interferes with the primary field generated by the probe.

If a flaw is present in the component, the eddy current will be disrupted with a consequential disruption in the induced magnetic field. This is detected by its effect on the primary field.

Eddy current testing relies on the skill of the operator to evaluate the signal changes, and the necessary level of skill and training required is higher than for other techniques. The degree of defect resolution with eddy current testing is comparable with that of magnetic particle or dye penetrant testing. However, it has several advantages over these other tests:

- it can detect subsurface defects
- the test produces immediate results
- the probe does not need to touch the component
- it is a clean test involving no magnetic liquids, dyes etc
- it can be used for more than flaw detection
- it can be used to inspect complex shapes and sizes of conductive materials
- minimum preparation of the component is required.

The relevant British Standard for eddy current testing is BS EN 1711 (BSI, 2000).

## Flaw detection by ultrasonic testing

Ultrasonic testing is a non-destructive test technique using high frequency sound energy to detect internal and surface breaking flaws. It can also be used for thickness measurement and material characterisation.

In an ultrasonic test, short pulses of very high frequency sound energy, typically in the range 1 MHz to 100 MHz, are introduced into the component. Any reflected sound pulses are then captured and analysed.

Where there is a discontinuity in the path of the sound wave, such as a crack, part of the wave will be reflected back from the surface of the flaw. The reflected wave signal is transformed into an electrical signal by a transducer and displayed on a screen. The wave travel time is related to the distance that the signal has travelled and information about the flaw's size and location can also be obtained from the signal.

The resolution of this technique depends on the frequency of the signal but although under laboratory conditions defects as small as 0.04 mm can be detected, in the field a practical limit in fine grained steel is a single millimetre sized defect.

The main disadvantages of ultrasonic testing are:

- cast iron is difficult to inspect due to low sound transmission and high signal noise due to the presence of graphite flakes. Older cast iron can prove impossible
- the level of skill and training required of the operator is more than with some other methods
- a coupling medium to allow transfer of sound energy into the test specimen is usually required
- rough, irregular, very small, exceptionally thin or non-homogeneous components can prove difficult to test
- linear defects oriented parallel to the sound beam may not be detected
- reference specimens are required for both equipment calibration, and characterisation of flaws.

The relevant British Standards for ultrasonic testing are BS EN 1712 (BSI, 1997) and BS EN 1713 (BSI, 1998b).

## 6.2.4   Destructive tests

Methods for removal of samples for destructive testing are discussed in Section 6.2.5.

### Tensile Test

Where the approximate method utilising field hardness testing is not accurate enough it may be necessary to conduct a direct tensile test on a specimen removed from the component in question.

The specimen should be taken in the direction of principal stress in the component, which is usually parallel to the member axis. The specimen should preferably sample as much of the thickness as is practical, depending on the accuracy of measurement required, as the properties may vary through thickness. This applies particularly to cast and wrought iron where the quality is likely to be more variable on a micro scale. In thick materials more than one test piece can be extracted within the depth, if necessary.

The size of test piece will be limited by the amount of material that can safely be removed from the member without impairing its structural integrity.

There are various standard sizes of test piece, either rectangular or circular in cross-section (see Figure 6.2), but the underlying principle is the same for all of them. The coupon is machined to produce a smaller cross-section in the middle (the gauge length) and a larger cross-section at the ends where it is gripped in the testing machine. The purpose of the gauge length is to ensure that the specimen fails in pure tension, unaffected by the grips.

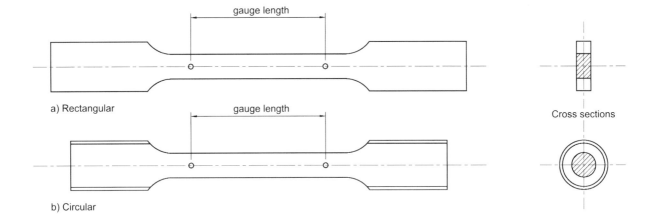

**Figure 6.2**      *Tensile test specimen*

BS EN 10002-1 (BSI, 2001) specifies a minimum gauge length of 20 mm for a standard geometry. This would be appropriate for a 4 mm diameter parallel length and would require a minimum total test piece length of about 60 mm in practice, depending on the grips and the size of the extensometer.

An example of a tensile fracture of a wrought iron test piece is shown in Figure 6.3.

**Figure 6.3**        *Tensile fracture of wrought iron*

For general fitness-for-purpose, where evidence of strict compliance with a product standard is not being sought, it is reasonable to use a shorter non-proportional gauge length of about three diameters (instead of five diameters) without affecting the yield or proof strength measurements. Extensometers are available for shorter gauge lengths than 20 mm. Elongation and tensile strength values are only likely to be affected marginally, if at all. This enables a 4 mm diameter test piece to be extracted from a 25 mm diameter cylindrical core (see Figure 6.4). The full length of the test piece can be obtained by welding steel extension pieces to the grip section. Electron beam welding should be used to ensure that the heat affected zone does not affect the properties of the parallel section. In material 20 mm thick two test pieces can be obtained from the depth.

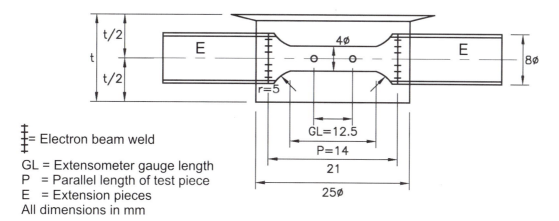

‡ = Electron beam weld

GL = Extensometer gauge length
P  = Parallel length of test piece
E  = Extension pieces
All dimensions in mm

**Figure 6.4**        *Extraction of a test specimen from a 25 mm diameter core*

The load-extension curve enables the yield strengths (or proof strength), elongation and tensile strength to be obtained. It should be noted that the tensile strength of wrought iron is very variable. BD 21/01 (HA, 2001) and Bussell (1997) both recommend characteristic values of 220 N/mm² which may be used without material testing if the metal appears to be sound. BD 21 recommends 230 N/mm² for pre-1955 steel and 46 N/mm² (as a working stress) for cast iron under permanent loads.

If testing is required then Annex C to BD 21/01 gives recommendations on the procedure to follow. However, it notes that testing only a few samples is unlikely to give a more reliable value than 220 N/mm² for wrought iron as this value is the result of a large number of tests.

Cast iron is also amenable to tensile testing, but being a brittle material the results are subject to scatter. In order to obtain a safe working stress it is necessary to test multiple samples. Statistical analyses can then be used to determine a safe load for a required probability of failure.

## Charpy impact test

The Charpy impact test is a destructive test used to measure the energy absorbed by a standard specimen impacted by a knife-edged pendulum. In practical terms it can be considered to measure the toughness of the material. In the test a coupon of the material to be tested is removed and then machined to a standard size of 10 mm × 10 mm × 55 mm with a 2 mm deep V-shaped notch at mid-point of the longer dimension.

The axis of the test piece should always be orientated parallel to the direction of main tensile stress in the member being sampled. The notch has a standard shape and tip radius and should be orientated in the through thickness direction to ensure that as many layers of steel are sampled as possible.

Small size test pieces having widths of 7.5 and 5 mm may be used for material thicknesses less than about 11 mm.

The specimen is fractured by striking it with a pendulum of known mass which is allowed to fall from a known height. The pendulum strikes the specimen directly behind the notch and the amount by which the pendulum swings through the vertical after impact is a measure of the energy absorbed in fracturing the specimen. The higher the pendulum swings the less energy has been absorbed and the lower the toughness of the material.

The test is usually carried out at the minimum design temperature for the bridge. A minimum of three samples is usually taken from each component, as the results can be quite variable. In exceptional cases a transition curve may be required, in which case up to ten test pieces may be needed and these will be tested at a range of temperatures to obtain the brittle-ductile transition temperature.

Charpy testing is not routinely undertaken because, for existing bridges, material toughness does not usually govern the assessment unlike, strength. The exception to this is, as mentioned above, where welded repairs are contemplated.

The relevant British Standard for the Charpy test is BS EN 10045-1 (BSI, 1990).

## Fracture toughness test

This is a very specialist test. It would normally be required if there was a particular concern about the risk of brittle fracture. This may arise in the case of bridges where evidence of fatigue cracking in tension members had been observed, where serious fabrication defects had been identified or where weld repair was required. If the Charpy test indicates a borderline situation, then the fracture toughness test can provide more definite data which can be used directly for an engineering critical assessment (ECA) of the risk with reference to BS 7910 (BSI, 2005).

The test piece is generally larger than in the Charpy test and has a sharp notch instead of a radiused V-notch (see Figure.6.5). The test piece is amenable to having welded extensions to small cores.

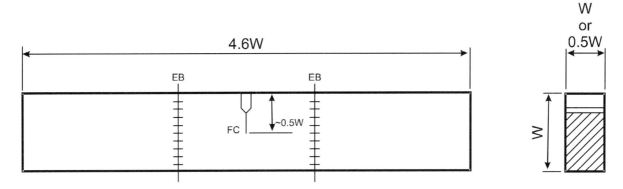

EB = Optional beam weld position
FC = Fatigue pre-crack

**Figure 6.5**       *Fracture toughness test specimen*

Selection of location, size and orientation of samples and analysis of results should not be done without prior consultation with specialists in fracture mechanics testing.

## Fatigue crack growth test

This is a very specialist test. It would normally be required if evidence of fatigue cracking had been found on a bridge and there was a concern that there could be other locations where fatigue had already initiated, but where the crack had not yet propagated to a size which was readily detectable by NDT.

In this case predictions of crack growth rate would be needed in order to develop a strategy for future inspections as part of a mitigating safety case to keep the bridge in service. The value of obtaining crack growth data from the material in question is that future inspections are likely to be less frequent than if upper bound data are used. The methodology in BS 7910 is normally used for this purpose.

The test piece is similar to the fracture toughness test piece but usually with a shallow notch (see Figure 6.6), and can be made from small cores by welding.

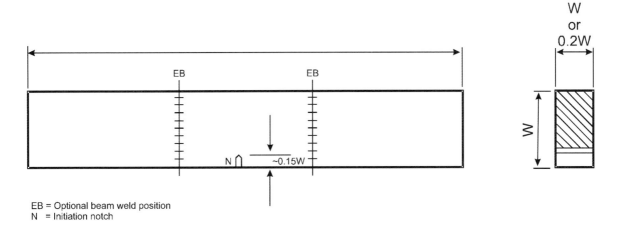

EB = Optional beam weld position
N  = Initiation notch

**Figure 6.6**       *Fatigue crack growth specimen*

## Compression test

The compression test specimen is a simple machined cylinder subjected to a compressive load between parallel hardened anvils of a test machine. The standard compression test for metals is BS EN 24506 (BSI, 2003). The advantage of this test is that proof strength can be estimated using a shorter test piece than needed for a tensile test (see Figure 6.7). Strain is usually measured with electric resistance strain gauges. The crushing strength of cast iron can be obtained from this test.

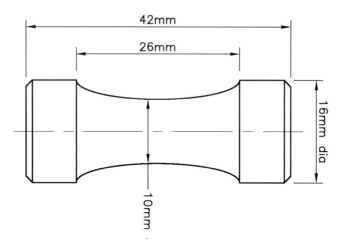

**Figure 6.7**     *Compression test specimen*

## Short transverse reduction of area (STRA) test

The STRA properties of steel are obtained by testing to BS EN 10164 (BSI, 2004). These properties are of particular interest when repair or strengthening measures involve new welded cruciform or T-joints onto old steels. The older steels tend to be higher in impurities, particularly sulphides. This can give rise to lamellar tearing in the parent metal either in or close to the heat-affected zone, which can severely weaken the joint when tensile stresses are applied through the thickness of the original steel.

Figure 6.8 shows the shape of the test piece, which is easily cored out of the component and machined to profile for testing. Unless the thickness exceeds about 40 mm it will be necessary to extend the core by electron beam or friction welding.

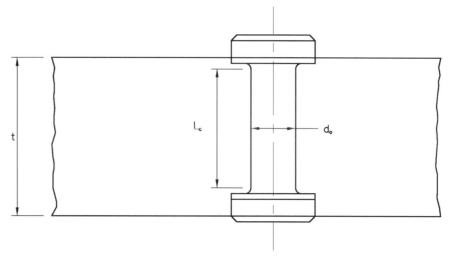

**Figure 6.8**     *STRA test specimen*

The test piece is strained plastically in tension until fracture. The reduction in cross-sectional area of the fracture is measured and expressed as a percentage of the original cross-sectional area. This is termed the "Z" value. Values of about Z15 give reasonable protection against lamellar tearing. Very heavy butt welds might require Z25 or Z35 to give complete protection, but older steels may not meet these properties.

## Micro-examination and fractography

The purpose of carrying out a micro-examination is usually to obtain evidence to explain some service malfunction. This can either be due to flaws in the original material or introduced during fabrication (in welds in particular) or in-service cracking or fracture.

In the case of flaws, the microstructure can give clues. This usually involves high magnification examination of micro-sections. The material is prepared to a higher level than for macro-examination. Evidence is usually obtained from extracted samples. However, it is possible to prepare replicas of exposed surfaces on site. These are prepared in the same way as it in the laboratory, but entirely by hand, followed by application of an acetate film. This is allowed to set, after which it is peeled off and taken back to the laboratory where the microstructure film is sputter-coated with a heavy metal and can be examined under the microscope.

In the case of fractures or fatigue cracking the fracture faces should be extracted with minimum damage and should always include the fracture tip region intact. The surfaces should be protected from corrosion as far as possible. Examination of a clean fracture surface can often yield important information about the mode of fracture and, in the case of fatigue, the crack growth history.

Micro-examination and fractography require special expertise and should be carried out by a materials specialist.

## Residual stress measurement

There are various methods for estimating the level of stress existing in a structural member. This will be the sum of any applied stress (dead load plus live load) plus any residual stress arising from the manufacturing process. The latter can be the result of:

- rolling of the original material and differential cooling
- weld shrinkage during fabrication
- lack of fit of parts during sub-assembly or erection.

In some instances residual stress measurement can be used to confirm whether or not weldments have been subject to thermal stress relief treatment.

The least destructive method of measuring such stresses is to use the centre hole air-abrasive technique. This involves attaching a special triple arm strain gauge to the metal surface and recording the zero reading. A small hole approximately 2 mm diameter and 2 mm deep is abraded into the metal at the centre of the gauge using a blast of fine powder. This relaxes the surrounding material depending on the earlier state of stress in the removed material (see Figure 6.9). This is measured by the change of strain in the three strain gauge arms. The principal stresses and directions can be obtained. Usually the hole can either be left or ground away to a smooth profile. The method only measures the state of stress at the surface local to the hole, so a number of readings may be required around a section.

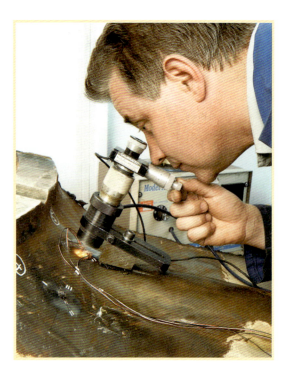

**Figure 6.9**

*The centre hole air-abrasive technique*

A variation to this principle is to trepan the surface using a coring tool which leaves a circular column of material which becomes relieved of residual stress. The change of strain may be measured by a strain gauge or extensometer. If the hole is trepanned through the whole thickness, measurements can be made on both outer surfaces of the resulting core. This method provides a better average residual strain estimate through the depth but is more invasive.

This principle can be made more global in its measurement by extracting larger sections of material with multiple strain gauges attached. It is important that the machining is done with sharp tools and no heat build-up.

A proprietary technique called alternating current stress measurement (ACSM) has been under development for measuring surface stresses in ferromagnetic materials. It involves use of a special probe which has to be calibrated on extracted samples of the same material which are subjected to known applied stress in the laboratory. The technique cannot readily distinguish between residual stresses and applied dead load stress. It is costly and requires a specially trained operator. For these reasons the number of bridges on which it could potentially be used is limited.

### 6.2.5 Extraction of samples

#### General principles

The three main methods of sample extraction are machining (coring, sawing, drilling), abrasive cutting and thermal cutting. The main differences are accuracy and metallurgical damage through heat build-up in the sample. Other considerations are limitations of access for the equipment, loss of stress carrying material, introduction of notches, ease of repair (if required), cost and time.

Mechanical methods are more precise, leave a better surface finish with fewer notches, and do not run the risk of alteration of the properties of the sample material through the build-up of heat. For these reasons mechanical extraction is preferred, particularly where the amount of material permitted to be lost is restricted. This factor may make the difference between being able to leave the excavation untreated or having to replace the loss of section.

It is essential that samples are extracted only from lightly loaded parts of the structure. If necessary this should be confirmed by analysis before removal.

### Coring

The least invasive method is by coring. It is quick and in many cases, if the core locations are selected carefully, the loss of section will be small enough not to require replacement. Empty round holes do not normally represent an unacceptable stress raiser. In the case of riveted girders, if they are located between lines of rivets they will not reduce the net section more than the rivet holes. If corrosion or ingress of water (for example in the case of box girders) is a potential problem, the holes can be plugged with plastic inserts or nuts and bolts, or tapped and screws inserted.

Cores can be taken from plain material or welded joints (see Figure 6.10 for various options). Hole sizes can vary from about 15 to 20 mm in practice. Cores as small as 10 mm may be suitable for chemical analysis, hardness testing, macro-sections etc. If extension pieces are to be used then cores of 21 to 25 mm diameter will be adequate for most of the mechanical tests described in Section 6.2.4, the hole sizes being about 30 to 33 mm in diameter. Note that extension pieces will not be suitable for cast iron. The smaller sizes of most mechanical test pieces can be extracted in full from a 75 mm diameter core.

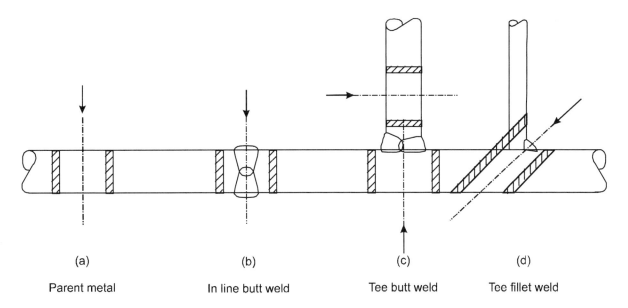

|  (a)  |  (b)  |  (c)  |  (d)  |
| Parent metal | In line butt weld | Tee butt weld | Tee fillet weld |

**Figure 6.10**    *Typical cores that can be removed from welded joints*

### Sawing

Handheld power saws have very limited application on their own but can be useful in conjunction with drilling.

There is a fully mechanised tool for extracting boat-shaped samples from the surface of a member. This is particularly useful where the material has to be fully restored by welding, as the excavation is shaped to provide ready deposition of weld metal (see Figure 6.11).

**Figure 6.11**

*Extraction of boat shaped samples*

The equipment is mounted on the surface with temporary welded or tapped studs. It has special spherically shaped saw blades which cut into the surface at the appropriate angles. Depths of cut range from 10 to 30 mm and lengths from 70 to 110 mm. Examples of boat shaped samples are shown in Figure 6.12.

**Figure 6.12**

*Extracted boat shaped samples*

## Drilling

Larger sections of any shape can be cut out by stitch drilling. For handheld power drills a maximum practical diameter is about 13 mm, smaller diameter holes being more efficient. This is a time consuming operation. For holes larger than 13 mm, a coring machine would be more suitable. Use of a template can help, particularly in keeping the holes close together. However , they should not intersect (see Figure 6.13). Final separation of the sample is usually by hacksaw or jigsaw.

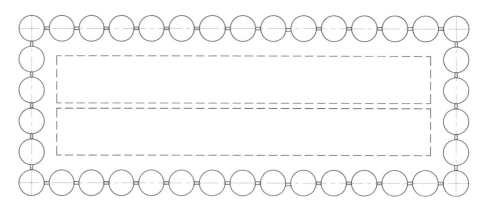

**Figure 6.13**          *Removal of a sample by stitch drilling*

### Abrasive cutting

High speed disk cutters can be used for straight cuts only. They can be useful for removal of angled samples from exposed corners eg the unstressed ends of I-girder flanges (see Figure 6.14). Any other geometries leave problems of severe notches. The equipment and operators are usually readily available, so that samples can be extracted without much delay in such a situation.

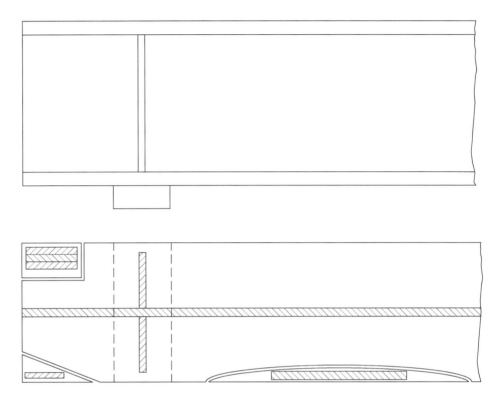

**Figure 6.14**     *Typical applications of abrasive cutting*

### Thermal cutting

This can be used for extraction of through thickness samples of any required shape. Again the equipment and operators are easy to obtain at short notice. In stressed regions the cut can be shaped to eliminate severe stress raisers. In tension members it is always advisable to grind off the hardened layer afterwards to remove notches or microcracks. At least 6 mm should be allowed at each cut for inaccuracy plus the effect of heat on the sample edge.

## 6.2.6     Load testing

It is known that the vast majority of bridges have real strengths significantly greater than conventional assessment predicts. There are two main reasons for this. Firstly, to ensure that designs are safe, design codes have a number of conservatisms built in to them. Factors of safety are applied both to design loads (which are themselves almost impossible to achieve in practice on anything but the longest span structures) and to material strengths. Secondly the analyses used in design are almost always simplified linear elastic analyses which ignore non-linear effects and the secondary load paths which invariably exist in real structures.

The upshot of this is that many structures which fail a conventional assessment will in fact be stronger than the assessment predicts. Load testing provides one possible means of identifying the hidden strength of these bridges.

## Static testing

Static load tests can be divided into two basic types:

1    Supplementary load tests.
2    Proof load tests.

A supplementary load test is a test designed to calibrate an existing computer model, or to confirm a suspected load path through a structure. A supplementary load test by itself cannot determine the capacity of the structure but should be considered part of the overall assessment process.

The level of load used in a supplementary load test should be sufficient to produce a measurable effect on the bridge without causing any permanent damage. Usually such loads are no more than experienced by the bridge under normal traffic loading.

An example of a supplementary load test was the test performed on Coalport Bridge in 2002, described in Appendix A2.1. This bridge, originally constructed in 1799 and believed to be the oldest cast iron bridge still open to vehicular traffic was assessed by its joint owners, Shropshire County Council and the Borough of Telford and Wrekin as having zero live load capacity.

During the design of the strengthening scheme it became apparent that the behaviour of the bridge is very sensitive to the support conditions, particularly those at the springing points of the ribs. A number of strain gauges were attached to the bridge to measure its response as an accurately weighed vehicle was driven across. The values of the support stiffnesses in the model were then adjusted to match the behaviour of the real structure.

Further in-depth advice on supplementary load testing can be found in the publication *Guidelines for the supplementary load testing of bridges* (Institution of Civil Engineers, 1998).

A proof load test is a test designed to demonstrate that a structure has a specific load carrying capacity. Unlike a supplementary load test, however, a proof test gives a load capacity without a need for further analysis. In a proof test the load is increased incrementally until it either reaches a predetermined value or the bridge starts to show signs of distress. The safe load capacity is then determined by dividing this value of load by an appropriate factor of safety.

From this it will be immediately apparent that the actual loads applied during a proof load test can be significantly greater than for a supplementary load test. Consequently, the risks to the structure are proportionately greater. For this reason proof load testing is not recommended.

Load testing in any form is expensive, usually requiring closure of the structure, design and installation of instrumentation to measure the response, and expertise to analyse and interpret the results. Nevertheless because it involves the real structure it can produce information about its behaviour which is impossible to obtain by other means.

## Dynamic load testing

Dynamic load testing can be used to determine values of stiffness to calibrate a computer model, and the dynamic properties of the structure. The latter include damping values and vibration frequencies. Values of the dynamic characteristics are required in situations when it is suspected that repair or strengthening work may cause the structure to become unduly lively when excited by traffic, pedestrians or wind.

There are several methods of applying dynamic load, all having been used successfully at one time or another:

- the most commonly used is vehicular loading where a loaded truck is driven across the bridge. A plank is fixed transversely to the direction of travel so that the vehicle bounces and applies a dynamic load after it is driven across. The response of the bridge can be recorded using LVDTs measuring deflections between the deck and ground, or accelerometers. Analysis of the response can be complex as the vehicle will apply a series of impulses that can interact with the response of the deck

- a dynamic exciter positioned on the deck at mid-span can be used to apply excitation at a predetermined frequency, usually the first mode bending frequency. A weight is oscillated in a vertical plane by hydraulic jacking where the frequency can be tuned to match the relevant bridge frequency

- a weight can be underslung from the bridge and dropped onto a safe place below. When attached to the bridge at mid-span the weight causes the deck to be deflected by a measured value that can be used to check its calculated stiffness. On release of the weight, the deck springs upwards and vibrates at its first bending frequency

- a weight can be suspended at a predetermined height and dropped onto the deck from above. It is of course necessary to protect the deck by a wooden cribbage. The drop-height is calculated to equal a half-cycle at first bending frequency. Using this method, the excitation is twice the value applied by simply dropping a weight off the deck

- small rockets can be used to apply impulsive forces for predetermined time-periods

- footbridges can be disturbed by a pedestrian or more than one pedestrian walking in step at first bending frequency at a rate aided by a bleeper. The bleeper can initially be set at the calculated value of bending or sway frequency and subsequently adjusted to the actual value.

Choice of the preferred method is usually a matter of convenience and available resources. For a bridge with a relatively short span, the quickest and cheapest method is with a loaded truck. For all spans a dynamic exciter will provide the best data. For relatively long spans, dropped weights are likely to be more practical.

## 6.3 MONITORING

### 6.3.1 Fatigue monitoring

The types of section usually used in bridge construction and, more specifically the connection details, dictate that fatigue failures usually initiate at welded joints, bolt or rivet holes and at sharp changes in section. Characterisation of joint types by their susceptibility to fatigue is possible. However, for all practical purposes, detection of incipient fatigue cracks during the initiation phase is not possible.

Once a fatigue crack has initiated it may be detected, if not visually, by any of the methods described in Section 6.2.3. Once detected the engineer will need to know how rapidly the crack is growing so that a decision can be made on whether immediate action is required, the bridge should be subject to a weight restriction or be closed immediately.

If the size of the crack and its rate of growth are known, then standard fracture mechanics techniques may be used to determine its effect on the structure.

A useful tool for predicting fatigue crack growth is the fatigue fuse, shown in Figure 6.15. The fatigue fuse is attached to the component to be monitored and is sufficiently thin and flexible that it does not affect its structural behaviour. The fuse contains a sharp edged slot which acts as a crack initiator. Once attached, cyclic loading of the component initiates cracking at the tips of the slot. As these cracks grow they fracture the conducting strips printed on the surface of the fuse. These fractures are recorded by the instrumentation attached to the fuse.

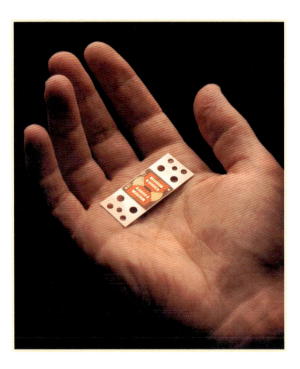

**Figure 6.15**        *A fatigue fuse*

Measuring the rate at which the crack in the fuse grows, together with an analysis of the structure, allows the engineer to predict the future behaviour. However, it must be appreciated that the fuse only begins to count the load cycles once it is attached. For structures which are well into their service life all the information up to the point at which the fuse is attached is inaccessible.

## 6.3.2     Acoustic monitoring

When a metal component is loaded, elastic strain energy is stored. If the component fails this energy is released. In certain circumstances acoustic sensors attached to the component can detect this failure and multiple sensors can be used to locate it. This is the principle behind acoustic monitoring.

A common use for acoustic monitoring is on the cables of suspended bridges. These have two principal features which make them good candidates for acoustic monitoring:

1    They are very heavily loaded so the energy release from broken wires is significant.

2    They are essentially one-dimensional systems, making location of the break relatively straightforward using data from only a few sensors.

As a result of these factors the signal-to-noise ratio in monitoring a metal bridge will be much lower than in simple system like a prestressing cable. In practice this means that using acoustic monitoring in metal bridges requires a great deal more skill in conditioning and interpreting the data, and the results are likely to be less definitive than for cables. On balance, acoustic emission is best applied to cables (suspension, stays or hangers).

## 6.3.3     Static deflection monitoring

Comparing the deflection of a real bridge with its computer simulation is one method of calibrating a computer model. However, highway bridges are usually enormously stiff structures and measuring deflections is not an easy undertaking. Possible techniques are the use of dial gauges or linear variable differential transformers LVDTs, photogrammetry and laser scanning.

The use of dial gauges or LVDTs is often impractical as they require a fixed reference point against which to measure the deflection. This may mean either construction of a temporary structure under the bridge or running a wire (usually Invar to minimise temperature effects) to ground. In either case, even if it is possible to obtain a measurement, each gauge or LVDT gives the deflection at one position only.

Static deflections can also be measured using conventional surveying techniques.

Photogrammetry is the use of photography for surveying. There are two broad fields of photogrammetry:

1    Interpretive.

2    Metric.

Metric photogrammetry is the branch which deals with the use of photographs to precisely measure the geometry of an object and is the branch relevant to static deflection monitoring of bridges.

A photographic image is a 2D projection of a 3D object. The basic principle of metric photogrammetry is to determine the relationship between the position of points in the 2D image and the position of the corresponding points on the object. Once this relationship has been established measurement of the object can proceed solely from the photograph by trigonometry.

The advantage of photogrammetry is that once the photographs have been taken many measurements may then be made corresponding to all of the parts of the structure within the field of view of the camera.

Laser scanning generates a 3D point cloud of the surface of the structure which can be processed to produce basic geometric information. This can be used either as the basis for a computer model or measuring deflections.

There are a number of commercial firms which offer a laser scanning service. The basic apparatus comprises the scanning laser which uses sweeping or rotating mirrors to pass a laser beam from a stationary source over the object to be scanned, and the post processing software.

While the accuracy of these systems is improving, and is certainly sufficient for geometric modelling, it is not nearly as precise as the other methods discussed above.

# 7 Structural assessment

> This chapter deals with the different levels and methods of structural analysis relevant to iron and steel bridges. The subjects addressed include:
> * trusses and girders
> * section strength
> * connections
> * composite action
> * fatigue.

## 7.1 INTRODUCTION

### 7.1.1 Object

The primary objective of assessment is to check that a structure is safe. Assessments are undertaken routinely to check that bridges are safe under the loads they are already experiencing, and also when loading is to be increased or the bridge is to be modified in some way. Bridges are also sometimes assessed when it is planned to drive abnormally heavy loads across them. Another important reason for assessing structures is when damage or deterioration has been identified and assessment is needed to see if the structure remains safe with the damage. Often, even when it is immediately obvious that remedial action is required, it is still necessary to assess the structure in its damaged condition to see if it is safe to leave it in service or even standing at all in the damaged condition.

### 7.1.2 Treatment of damage and deterioration

Although BD 21/01 (HA, 2001) allows for the approach of assessing the structure in its undamaged as-built condition and then applying an overall condition factor to allow for deterioration, this approach is unlikely to be appropriate in the case of a damaged bridge. It is normally necessary to quantify the damage to particular elements and then allow for it explicitly in calculations. Specific ways that damage can be considered will be discussed in later sections of this chapter.

### 7.1.3 Codes of practice, philosophy and differences from design

The Highways Agency and Network Rail have issued special assessment versions of their main design codes, covering loading and structure strength assessment. The main difference between assessment and design loading are:

* emphasis on ultimate strength

* less conservative loading eg removal of 10 per cent contingency factor for HA design load

* coverage of reduced loading to enable weight (and also, in the case of rail, speed) restrictions to be determined for bridges that are not adequate for the full load.

The assessment codes also aim to remove some of the conservatism from design codes.

Another issue is that design codes represent what is currently considered good practice. There is no reason for them to consider details or forms of construction which have been superseded. In contrast, assessment codes would ideally cover all forms of construction that have been used. The Highways Agency's steel and composite assessment codes, BA 56/96 (HA, 1996) and BA 61/96 (HA, 1996), attempt to address this by covering a wide range of cases but inevitably there is still not full coverage. The result is that more judgement is needed in assessing old structures than in checking the design of new ones. It is also worth being aware that the Network Rail Underbridge Code (Network Rail, 2004b) also gives additional coverage which may also be useful for bridges carrying things other than railways. There is also a London Underground Assessment Standard (London Underground, 2006) which gives useful guidance.

Although these special assessment versions of the main bridge design codes have been produced, they are based on design codes which have a much longer history. The assessment of bridges, like other structures, is still usually done using approaches which were originally developed for design. This seems logical, assessment is superficially similar to design, or at least to checking new designs. However, there are important differences between the design and assessment situation which should fundamentally alter the approach. If this is not appreciated, it can result in a significant waste of resources on unnecessarily sophisticated analysis. More seriously, it can result in structures being strengthened or even condemned when they are in fact satisfactory.

The analytical approaches used in design are often conservative. Alternative approaches are available which give more realistic results. These approaches are often, although not always, more expensive. The high cost of strengthening existing structures makes it more likely to be worth using them in assessment. However, while the designer aims for a consistent margin of safety, and accordingly uses consistent analytical approaches, the margin of safety in existing bridges may vary greatly. As a result, there are also cases where a very simple conservative analytical approach is adequate to show that a structure is satisfactory, so a much wider range of analytical approaches are used in assessment than in normal design.

In general, the best final assessment approach is the cheapest method which shows the structure to have the required strength. Since the assessed strength will not be known until the assessment is well advanced, it will sometimes be necessary to refine the approach and use more advanced analysis or possibly more accurate (less conservative) determination of damage, material properties or geometry and repeat part of the process as illustrated in Figure 7.1. Reviewing the results is an important part of the process. It is reasonable, indeed desirable, to assess bridges as safe using simple conservative approaches. However, it is never appropriate to condemn them on the basis of such assessments without first reviewing the findings and asking the question "is this failure real, or merely the result of a conservative assessment?" The result is that assessment can be an iterative process involving progressive screening, using more sophisticated and realistic approaches until either the bridge is found to be adequate or it is concluded that shortfalls identified by the assessment are real and some action is required.

In the following sections, a wide range of assessment techniques are considered from simple rules given by documents such as BD 21 /01 (HA, 2001) and BA 16/97 (HA, 1997) to sophisticated non-linear analyses. References are given to specific assessment codes where appropriate, but clauses and values are not reproduced. The documents to be used are often specified by bridge owners, and engineers will have to refer to the appropriate ones. Under formal procedures such as used by the Highways Agency and Network Rail, some of the more advanced approaches considered will require departures from standard. However, many will not. Even when departures are needed, the assessments are not totally outside the codes; and the philosophy, safety factors and material properties, and other rules from the codes are still used, albeit sometimes with modifications. It is also often helpful to refer to assessment codes other than those specified, as different codes may give better coverage of, for example, particular forms of construction. However, if code clauses from other codes are used, it is necessary to check that the safety factors are consistent and modify them if not.

### 7.1.4      Management of under-strength bridges

When a bridge is assessed as inadequate for the full assessment load, a number of options are available but enhancement of the assessed capacity is preferred where possible, as indicated in Figure 7.1. Even very sophisticated analysis and extensive testing is normally cheap compared with strengthening. Sometimes, however, the cost of the most refined analysis or the cost and other implications of testing to refine the properties used in the assessment is not considered justified. This can arise, for example, when it is considered that load restriction does not cause problems.

BD 79/06 (HA, 2006) covers the management of sub-standard structures during assessment including the important issue of what interim measures should be taken. BD 79/06 considers three levels of assessment:

**Level 1**    Assessment is simple analysis (see Section 7.3.1).

**Level 2**    Can include more sophisticated analysis considered in this chapter. However, once new material tests are taken, it counts as a Level 3 assessment.

**Level 3**    Assessment can also include deriving bridge-specific live loading to BD 50/92 (HA, 1992) although this is only relevant for longer span structures with only relatively small shortfalls of strength.

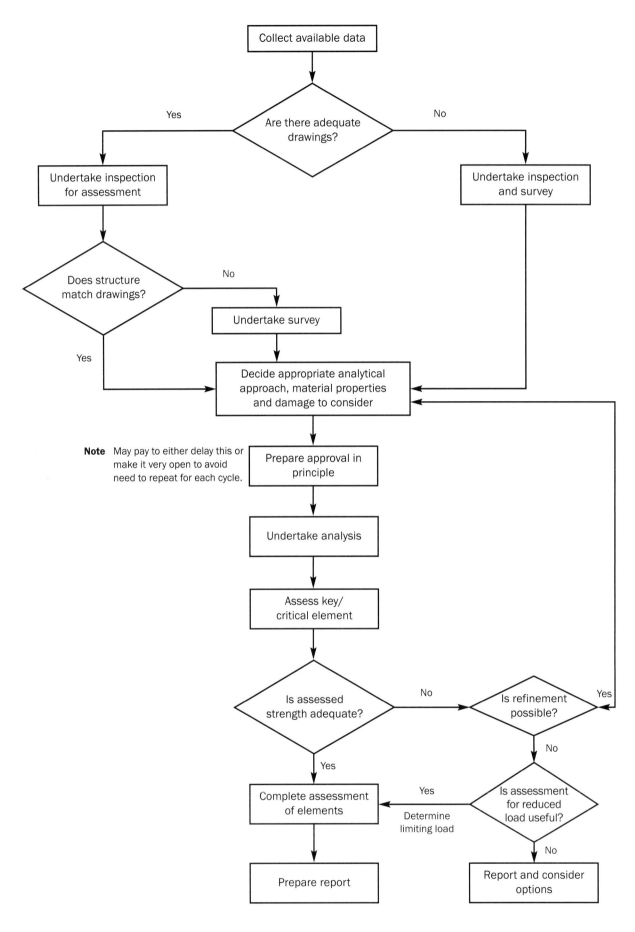

**Figure 7.1** *Assessment flow chart*

BD 79/06 (HA, 2006) notes that the division between the three levels of assessment is not clear. For example, with very simple beam structures, the benefits of more sophisticated analysis may be very small and a sensitivity analysis would show that more realistic estimates of material properties obtained from tests will be more beneficial, going directly to Level 3 in preference to, and before, Level 2.

BD 79/06 also raises the possibility of direct use of reliability based assessment methods if Levels 1 to 3 assessments have not shown adequate capacity. However, it does not elaborate and suggests such methods require specialist knowledge and expertise and, in effect, Category III checks. The occasions when this will be appropriate will be relatively rare and mainly on major structures, and the approach is not explicitly covered here or considered in Figure 7.1. It is mainly worth noting if there is a relatively small shortfall on a major structure. A particular case where it can be useful is if assessment shows a very low capacity (perhaps less than dead load) for a bridge which has been in service under full loads for some time. Although a bridge which has carried full 40 ton loading does not prove it is safe for this, Bayesian updating using this knowledge may enable it to be shown to be safe for less restrictive loading than a conventional deterministic assessment would indicate.

The options for strengthening and replacement are considered in Chapter 10.

Even if money is available for strengthening, it will rarely be possible to undertake such work immediately. Load restrictions are often needed in the interim.

The simplest form of load restriction is to apply a weight limit. However, these are notoriously difficult to enforce. It is not prudent to rely on them for cases where the bridge would collapse or suffer serious damage if the limit was exceeded. A more satisfactory form of weight restriction is to alter the road geometry to reduce the number of lanes. Where the required weight limit is very low, it is also possible to install physical measures to stop over-size vehicles getting onto the bridge. This is normally only practical for the lower weight limits required for very under-strength bridges because lorries of 17 and even seven tons are usually no narrower and often no lower than the largest standard lorries.

On rail structures, weight limits should be more reliable. Occasionally there is the additional option of imposing speed limits which can reduce the dynamic factor.

If the assessment really does show the bridge to be unsafe, it may be necessary to close the structure.

In principle, there is the additional option of monitoring. Some aspects of this are considered in Chapter 6. It can be very useful for structures that are assessed as having inadequate fatigue life, particularly when the critical details are damage tolerant. However, there is limited scope for using monitoring to keep in service bridges which have been assessed as having inadequate static strength. This is because the failure modes do not give suitable warning.

## 7.1.5 Financial liability

Many of the older metal bridges in this country carry public highways but are owned by bodies other than highway (or road) authorities. Part VIII of the Road Transport Act 1968 (HMSO, 1968) placed specific obligations in respect of these bridges where they crossed "a railway of the Railways Board, a railway of the London Board, an inland waterway of the Waterways Board …". Section 117 of the Act stipulates that the relevant bridges must have "the required load bearing capacity" and that this may be from an order by "the appropriate Minister" or, where no order is made, "the traffic which ordinarily uses or may reasonably be expected to use the highway carried by the bridge on or about the day on which this section comes into force" (ie 1 January 1969). The principal successor bodies to those listed in the Act are now Network Rail Infrastructure Ltd, BRB (Residuary) Ltd, London Underground Ltd and British Waterways. In addition these obligations may have passed to subsequent owners of such bridges, including light rail operators and preserved railways.

For railway bridges (defined as owned by the British Railways Board, the London Transport Executive, the National Freight Corporation or any of their subsidiaries) in England and Wales, the required load bearing capacity was fixed by SI 1705 Railway Bridges (Load Bearing Standards) (England and Wales) Order 1972 (HMSO, 1972). No similar order has ever been issued in respect of bridges in Scotland or those owned by British Waterways, for which the less detailed requirements of Section 117 will apply. The Order specifies different load bearing capacities for bridges of different ages and partly depends on the class of highway carried by the bridge. For bridges built before 1955 a calculation method is also specified in the Order. Although both vehicle and axle weights have increased since the legislation mentioned was enacted, the load bearing capacity specified by it has not been amended to reflect such changes.

This means that the owners of bridges covered by the Act have no legal (and consequently financial) responsibility for providing a full strength bridge where the capacity would be in excess of their obligations specified in the legislation. So for such bridges it is often necessary to undertake two separate assessment calculations. The first will be to the up-to-date code (usually BD 21/01), giving the currently accepted calculated capacity of the bridge. If this shows that the bridge is weak then a further assessment using the older methods will be undertaken (usually by or on behalf of the structure owner) to ascertain whether the bridge meets that owner's legal obligation. If this second assessment reveals a statutory failure then the bridge owner is obliged to contribute financially to any subsequent remedial action. If it shows a statutory pass then the highway (or road) authority is fully financially responsible for subsequent actions. It must be realised that the result of the second assessment (sometimes referred to as a Section 117 or BE4 (MoT, 1970) assessment) is merely a device for determining how financial responsibilities should be shared between the bridge owner and the highway (or road) authority and must not be used to replace the BD 21/01 capacity of the bridge or for determining restrictions on the use of the bridge.

## 7.2 INFORMATION REQUIRED

Before starting on assessment calculations, it is necessary to collect the information required. Normally an inspection for assessment will be undertaken and, if there are no adequate drawings available, this will include a full survey. Even if there are drawings, some key dimensions including thicknesses should be checked as it is not uncommon for structures to differ from the drawings. The survey will determine whether the structure can be assessed as-built or whether deterioration needs to be considered. Where the deterioration is due to corrosion loss, it is best to measure it. Subjective assessments of percentage loss of section tend to be very conservative.

It is also necessary to determine the material properties which will be used. Usually, if there are drawings these will indicate the materials used. If not, default values from BD 21/01 are usually taken, which can be quite conservative. However, due to the variability of early iron and steel, small numbers of tests are unlikely to give a reliable basis for increasing the assumed strength. Tests are normally undertaken at this stage only if the nature of the material is unknown or if there is reason to suspect the actual strength may be lower than would otherwise be assumed.

Since the assumed material properties are likely to be conservative, they are another aspect which should be reviewed if the initial assessed strength is inadequate. It is often worth doing a sensitivity analysis before undertaking tests to see if probable results achieve useful increases in assessed strength. If they do, it is possible to concentrate tests in the relevant area and type of material (for example, a particular plate thickness) making it more likely that a relatively small number of tests, say 6 to 10, can give a useful increase in the material properties used.

## 7.3 DISTRIBUTION ANALYSIS

Analysis is a major part of the assessment process as it is necessary to determine the forces that members have to resist to enable the bridge as a whole to resist the loads it is subjected to, before the safety of individual members can be assessed. Serviceability checks, which may be critical in design, are less significant in assessment of iron and steel bridges.

Theoretically, most of the analytical methods are not precise. They are theoretically safe only because of the safe theorem of plastic design. This means that as long as the elements are adequate for the assessed forces and the assessed forces are in equilibrium with the applied load, the structure will not collapse. This, however, depends on ductility. Cast iron is too brittle for this to apply and the special problems of cast iron will be considered in Section 7.8. The ductility of steel and wrought iron structures, particularly those that fail by buckling rather than yield, is also limited and this can restrict the analytical approaches that can reliably be used.

When sections have very localised corrosion or other section loss, the loss can be treated like rivet and bolt holes. Yield is checked using the net section allowing for the loss, but buckling can be checked using the gross-section. However the main types of analysis will now be considered.

### 7.3.1 Simple analysis

For simple bridges with steel or wrought iron girders, a simple static load distribution will normally give a conservative calculation of beam loads and a conservative strength assessment. There are exceptions to this with cast iron girders which will be considered in Section 7.8.

A static load distribution provides a set of forces which are in equilibrium, so it is a safe method according to plastic theory. If a structure has adequate strength according to such an analysis, it is normally safe. If only a strength assessment is required, there is usually no point in doing a more sophisticated analysis. The simple analysis also gives conservative figures for serviceability assessment of longitudinal members. However, it cannot provide any information about transverse moments since it ignores them. If serviceability assessment of transverse members (eg crack widths in a concrete deck slab of a composite bridge) is required, a more sophisticated analysis is needed. This may also be required, where transverse members are important to the restraint of the main members against buckling, to check the transverse members are not too extensively yielded to fulfil this function.

BA 16/97 provides a simple assessment method using empirical charts. This approach is more restricted in application than a static distribution and cannot be used for abnormal vehicle loading (eg HB or STGO) but it can be less conservative. There is also a similar approach for transverse members but this is still more restricted in application. Like the static load distribution, these methods do not give any information about the forces in transverse members and so cannot be used if serviceability assessment of these is needed.

The opposite extreme from a static load distribution is to assume the distribution is perfect and compare the total longitudinal moment and vertical shear on the bridge with the total available capacity. This is an unsafe method (it is in fact a particular yield-line mechanism) but it does give useful information. If a bridge fails this test, no amount of improved load distribution will make it work. It provides an upper bound to strength and also another useful check on other analyses. If a distribution analysis gives results which do not fall between those obtained using a static and perfect load distribution, it is wrong.

### 7.3.2 Conventional analysis

In the design of beam-and-slab type bridges, an elastic grillage or finite element (strictly, since grillage elements are finite, other finite element) analysis is usually used. This is also the most common method used in assessment and will normally give a better (more uniform) distribution of load between the girders and a greater assessed capacity than a static distribution. However, before undertaking such an analysis, it is worth noting its limitations.

The increase in strength comes only from improving the distribution of load between the girders. This distribution comes as a result of the transverse moments. In addition to the aforementioned fact that the distribution analysis cannot increase the total moment capacity of the bridge, this has another implication. If a bridge has no transverse moment capacity, a distribution analysis cannot help. Because many older bridges were designed using a static load distribution, they often had inadequate transverse strength compared with modern designs and a static load distribution may be the best way to assess them. It is quite common to find bridges that fail according to a conventional assessment due to inadequate transverse capacity when they pass according to an assessment based on a static load distribution.

Where all the members in the analysis are metal, they will be represented with elastic section properties. According to plastic theory the results would still be safe if different properties were used. However, this is not often done in metal structures and indeed girders that are not compact (ie slender so that failure can be by buckling rather than yielding) are not necessarily sufficiently ductile for plastic theory to be valid.

In composite bridges where the deck slab is of reinforced concrete there is more scope in the choice of assumed section properties. In design, it would be usual to use gross-concrete properties. This will give the best distribution but also the greatest transverse moments. In older bridges, which were designed using a static load distribution, the reinforcement will often be inadequate to resist these moments. In such cases it will be beneficial to reduce the transverse stiffness towards the cracked-elastic value.

Some bridges have cross-girders or bracing between the main girders which does not support the deck slab. This improves the distribution properties. However, they or their connections are sometimes overstressed according to the analysis, when the main girders are satisfactory. In such cases, it is worth removing the cross-girders from the analysis. The structure may still be adequate even if they fail in which case overstress is not important.

## 7.3.3　Other forms of analysis

It is relatively unusual to use plastic or non-linear analysis for determining the load distribution in metal bridges. In principle, redistribution can be used where the members are sufficiently ductile. However the requirements to enable a girder to redistribute moment without buckling are more severe than for enabling a plastic section analysis to be used. BS EN 1993 (2005) has different classes of section to allow for this but it does not normally depart from elastic analysis for bridges.

In principle it is possible to undertake a non-linear analysis which correctly allows for all the limitations on redistribution and this form of analysis is considered in Section 7.7. However, the requirements for doing this are quite severe. In particular, it is necessary to be able to correctly model all possible relevant buckling modes.

There are two cases where non-linear analysis is more likely to be beneficial.

1　One is in infill joist bridges. Here the concrete prevents premature buckling of the metal and the main non-linearities to be considered are in the concrete. This will be briefly considered in Section 7.9.2.

2　The other is in thin deck plates which are restrained in plan. Here, the deflections can be large relative to plate thickness so, if there is sufficient restraint, membrane action can greatly enhance the strength.

The considerations in the non-linear analysis of such cases are similar to those in Section 7.5. It is particularly important to consider the interaction with global effects. It is possible for a plate to resist out of plane loading eg from wheels by tensile membrane action at the same time as acting as the compression flange of either longitudinal or transverse girders. However, considering the effects separately can lead to unsafe answers.

## 7.4　ANALYSIS OF TRUSSES AND LATTICE GIRDERS

Where the main girders are simple beams, the assessment will proceed directly from distribution analysis to assessment of individual members. For truss bridges, or for truss and lattice girders forming part of bridges, it is necessary to analyse the truss to determine the forces in individual members making up the truss.

Until the advent of computer analysis, trusses were invariably analysed as pin jointed. Even after the general use of computers, it was common to analyse trusses in this way. Analysis taking the joints as fixed frequently predicts local moments in the members and their joints which they were not designed for. However, provided the behaviour is reasonably ductile, inability to resist these moments does not lead to failure. So for ultimate strength assessment, it may be preferable to analyse the joints as pinned. If this is done, it does not invalidate considering the joints as fixed for the purposes of evaluation of effective length for buckling. It may also be reasonable, for example, to consider the chords (which frequently have to resist moments due to slab or cross-girder loads applied directly to them) as continuous and only pin the connections to the bracing members.

Many older trusses had members which were only designed to take tension. They may have either been very slender (often flats or rods) or had connection details that work only in tension, or both. If a high live load is applied on, for example, only half the span, a conventional analysis may predict compression in such members. It is necessary to remove these forces to get realistic results. For simple cases, this can be done by hand by removing the offending elements from the computer model. However, with more complicated structures, this can become difficult particularly as different members may have to be removed for different load cases. Some programs have the facility to incorporate tension-only elements without invoking a full non-linear analysis and this is very convenient for analysing this type of problem.

There can also be elements which can take some compression but which appear to be overstressed due to being very slender. More realistic buckling analysis (considered in Sections 7.5.1 and 7.7.3) may solve this problem. However, an alternative worth considering for appropriate cases is again removing them altogether from the model.

## 7.5 SECTION STRENGTH ASSESSMENT

Having determined the forces in the members, it is next necessary to determine if the members can resist the forces. For simple compact beams this is done with plastic section analysis. If they are fully restrained, such as when they are embedded in concrete, the plastic section properties can always be used. For other steel and wrought iron structures buckling is a significant issue. Codes give semi-empirical rules for assessing the effect. Although these can sometimes get quite complicated, they are reasonably well covered by codes such as BD 56/96 (HA, 1996) and will not be considered in detail here. However, it should be appreciated that they are necessarily conservative. So when a structure fails the code rules but still has a reserve against yield, it is worth considering if the assessed capacity can be increased by more sophisticated analysis.

Occasionally, some sections may fail to comply with code values for such things as outstand ratios. It is reasonable to obtain a conservative strength assessment by assessing a hypothetical structure with the flange width reduced so that it complies. However, like all conservative approaches, this is reasonable for assessing a structure as safe but not for condemning it. If it fails to show the structure to be adequate, a more realistic assessment is required.

When sections have very localised corrosion or other section loss, the loss can be treated like rivet and bolt holes. Yield is checked using the net section allowing for the loss, but buckling can be checked using the gross-section. However, unlike bolt holes, corrosion is likely to be significantly asymmetric. In such cases, it is necessary to allow for stresses due to the reduced section not being concentric with the intact element either side of it.

Once the section loss is more extensive, it will be necessary to consider it in the buckling calculations. Since the code rules are largely based on uniform sections, this can only be done with simple code rules by making simple and usually conservative approximations to a uniform section. It is reasonable to approximate to a uniform section that is greater than the minimum, provided yield is checked everywhere using the actual local section.

### 7.5.1 Buckling analysis using Eigen values

Most modern programs used for linear-elastic analysis can give elastic buckling loads directly. These are obtained by determining the Eigen values. To do this properly, the computer model has to be fine enough to model the relevant buckling modes. This means, for example, that beams have to be represented with relatively fine mesh shell element models, not by simple beams.

It is sometimes possible to analyse for global buckling in one model and local buckling either by conventional hand methods or with a localised model. However, there are many cases where this is difficult because the effects interact. For example, bearing stiffeners are important to restraint of main girders. So analysing the global behaviour then checking the bearing stiffeners separately can be unsafe as it ignores restraint forces in the stiffeners.

It is possible to apply the approach to truss structures using line element space frame models. However, it is then necessary to have computer nodes between the physical joints otherwise the program cannot model buckling between the joints. It is also necessary to consider the behaviour of the joints carefully.

It should be appreciated that elastic buckling loads cannot be used directly for design or assessment. A simple elastic buckling analysis implies a specific buckling load with no second order effect before it is reached. However, the second order effect actually becomes significant earlier, increasing moments due to applied loading and also exaggerating the effect of imperfections. Although this does not reduce the elastic buckling load, it does increase the stresses under a given load, reducing the strength. Also, in reality, once the material yields the stiffness reduces and the second order effects increase so the buckling load reduces. Code formulae allow for this. If this form of analysis is to be used the approach is to replace the simplistic elastic buckling values, based on simplified structures in codes, with the more realistic value obtained from Eigen value analysis but still use code approaches after that.

The approach can potentially be used to analyse any of the possible buckling modes in steelwork and wrought iron. However, it is unlikely to be useful for analysing web buckling and codes do not make doing this very easy. It will also be found that the semi-empirical approaches in codes sometimes allow for post-buckling behaviour to some extent and can actually give strengths which are above the theoretical ideal elastic buckling mode.

The approach is not likely to be useful for analysing very simple cases such as a uniform pin-ended strut in compression. This is because the Eigen value buckling value will be the same as the code value. The approach is more useful for assessing bridges which for various reasons do not match the standard cases in codes. For example, if assessing lateral torsional buckling, or indeed strut buckling, in members with varying section or somewhat indeterminate restraint, by modelling the section and restraints thoroughly, a realistic elastic buckling load can be obtained.

If there is section variation due, for example to very local corrosion, it is reasonable to ignore it in the buckling analysis provided the reduced section is considered for yield. If it is more extensive so that it could affect the stiffness significantly, it should be considered in the analysis. In either case, if the loss is asymmetric, it will again be necessary to consider the effect of the eccentric load on the section. If it is asymmetric for a significant length, a second order effect of the resulting eccentricity could arise which this form of analysis cannot allow for and it is better to use a full non-linear analysis.

A particularly common case where bridges cannot be checked for buckling using conventional rules in current codes arises in bridges which rely on U-frame action. This is because the bearing stiffeners do not comply with the requirements or because the intermediate web stiffeners do not line up with the cross-girders. Accurate elastic buckling values for such cases can readily be obtained from the Eigen value approach.

There are some difficulties and limitations with the approach. Firstly, the modelled behaviour has to be linear-elastic. It cannot properly allow for supports that provide good rotational restraint initially but lift off later, or for joints that have wobble to take up before becoming effective. It can also be difficult to interpret the results particularly in terms of local checks. For example, the approach can determine the global capacity of a U-frame bridge that has support stiffeners that do not comply with stiffness requirements of current rules. However, there is no simple way of deciding how to check the strength of the support stiffeners. The approach can also pick up a web buckling mode when the web actually works by post-buckling behaviour. Although it may then be possible to check those panels by conventional code rules, it is not easy to determine if the relevant buckling mode would interact with others.

A further factor is that the major cost of undertaking a full non-linear buckling analysis compared with a simple analysis lies in putting together the necessary detailed model which is also needed for an Eigen vector analysis. However, while the non-linear analysis may require almost no hand post-processing, the Eigen value approach may require considerable work to check section strengths throughout the structure. The result is that, although the non-linear approach may require several orders of magnitude more computer time than the Eigen value approach, it may actually be cheaper.

## 7.6 CONNECTIONS

As well as assessing the basic members, it is necessary to check the connections. In addition to connections at joints, such as in trusses, there are often joints due to limitations on the length and thickness of plates that could be produced. When a thick plate is made up of multiple plates and cover plates, it may be that a discontinuity in one of the plates is not obvious from looking at either the drawings or the girder itself. Older structures did sometimes have quite short cover plates to increase thickness in critical sections. However, if the cover plates are very short, they are probably actually splice plates to compensate for a break in a plate that may not have been noticed and which may be four or five plates below the cover plate.

Older steel and nearly all wrought iron structures were riveted. Although rivets are covered in current assessment codes, they have not been used extensively since the 1950s. The design rules assume that the rivet strength can be simply derived from the yield strength of the parent material. However, in reality the rivets are stronger than the parent metal and significantly stronger than predicted by BS 5400. This was acknowledged by the more empirical rivet design approaches used in the past. This has been considered in work for Network Rail, and their Underbridge Code (Network Rail, 2004b) gives findings which are very useful.

Rivet holes can take up a significant part of the section. So as well as checking the rivets themselves it is necessary to check the net section of the element for yield.

Connections are sometimes encountered which incorporate both rivets (or occasionally bolts) and welds. In new structures this approach is not normally used, except when the bolts are only for the temporary condition. Because the welds are close to rigid, whereas the bolts and to a lesser extent rivets require significant movement to take load, the advice is normally to ignore the rivets and take all the load on the welds. An exception arises if the weld was added (as for strengthening) after the rivets were loaded by dead load. In this situation the dead load is clearly taken by the rivets and this can be considered in the assessment.

Except for very long joints, it is usually possible to use a plastic distribution of force between bolts, rivets or weld material in a joint. This does not always apply to long splices but the issues are covered by codes of practice. For flange to web joints, it is usual to use an elastic distribution of force along the joint. However if this predicts local overloading, it is possible to redistribute the force along the joint.

When forces are redistributed along a joint it is important to check that the material either side is still in equilibrium. This is not normally an issue for simple in-plane splices. However, very detailed finite element models of steel structures can make the issue important. Because of the stiffness of the web, the FE model would predict that the weld force would be very concentrated in the centre. It is perfectly reasonable to spread it along the weld but there would then be a local moment in the flange resulting from it cantilevering out from the web. This will not exist in the computer model and so has to be checked by hand.

Another issue that frequently arises is the extent to which compressive loads can be taken by direct bearing. Because of the importance of fatigue, modern bridge design codes are quite conservative on this. Unless the surfaces are machined, it is usual to assume none of the load is taken in bearing. With bolts and rivets, this is unduly conservative. Even with welds, static load can be taken in bearing where fatigue is not an issue.

## 7.7 NON-LINEAR ANALYSIS

### 7.7.1 Uses and approach

It was noted in Section 7.3.3 that non-linear analysis can, in principle, be used to undertake a distribution analysis which takes account of redistribution. However, except for infill-joist type bridges where buckling is restrained by the concrete, it is necessary to have a model that can cover all the possible buckling modes that might affect redistribution properties. In principle, it may also be necessary to consider the possibility that the behaviour of splices or other joints affect ductility, so this approach is rarely used.

Non-linear analysis is more often useful for investigating buckling problems. It has probably been used most commonly for analysing bridges which are dependant on U-frame action, either to obtain increased assessed capacity compared with conventional code assessment or because details of the structure do not comply with the rules and so conventional codes do not give a capacity.

Non-linear analysis can also be used to analyse more local problems such as with stiffeners or section strength assessment. BS 5400 considers sections as either compact or non-compact. The former are able to develop their full plastic moment capacity without buckling but the section capacity of the latter is checked elastically. This means that the code gives a sudden drop in capacity as the limit of compact sections is reached. For a simple universal beam, this is typically in the order of 10 to 15 per cent but for some other sections, it can be much greater. This sudden drop does not occur in reality and non-linear analysis can be used to give more realistic capacities for sections which marginally fail to comply with the criteria for compact sections.

### 7.7.2 Treatment of safety factors

The safety factor format in current design codes was developed for conventional analysis where elastic analysis is used and sections are checked separately. It is usual to use the nominal material properties in the elastic analysis and apply the material partial safety factors only in the section checks. In non-linear analysis, the section and structure analysis is done together and consistent properties have to be used. In steel structures where the material partial safety factor is both constant and small, it is often convenient to apply all the partial safety factors (including gf3 in BS 5400 and the assessment codes derived from it) to the loads. This is over-conservative particularly for buckling analysis as it is equivalent to applying the material partial safety factor to the E-value as well as the yield strength. However, it does have the advantage of being rigorously safe and simple.

With composite structures the concrete and steel have different material safety factors which makes it more convenient to apply the material partial safety factors to the material in the computer model. Since concrete stiffness, unlike that of steel, is related to strength this means the structure is less stiff than in reality. So the approach is potentially unsafe if imposed deformations (such as differential settlement or differential temperature) are critical. It is very unusual for such effects to be critical for ultimate strength in concrete structures, but it is more likely to be critical for the relatively brittle failure modes of composite structures that fail by buckling. There is no very straightforward solution to this issue and it is probably best to undertake a sensitivity analysis.

## 7.7.3     Analysis of buckling

The approach to buckling is to analyse the structure using a geometrically non-linear analysis which automatically takes account of buckling effects. As with the Eigen value analysis, the first requirement of a non-linear computer model to analyse buckling is a fine enough mesh to model all the relevant buckling modes.

If all the available restraints are included, a non-linear analysis will often give very much higher buckling loads than conventional analysis. It may be possible to get significant benefit without modelling all the restraint. It has been found, for example, that analysis of half-through girder bridges where only the lateral restraint at deck level is considered, not the U-frame action as such, can often give adequate strengths (Mehrkar-Asl *et al*, 2005). This avoids the need either to model the whole bridge or to determine the adequacy of the connections between cross-girder and stiffeners to provide U-frame restraint.

In order to obtain safe results from this form of analysis, it is necessary to consider the effect of imperfections. The standard way of doing this is to first undertake an Eigen value analysis to identify likely critical modes. Imperfections are then imposed on the model with the corresponding shape. Although most of the bridges assessed will predate BS 5400, the fabrication tolerances from that document are used as a guide and BD 56/96 (HA, 1996) recommends using half and twice the tolerances as a sensitivity check. The latter, particularly in global modes (eg lateral buckling of compression flange), usually gives such large imperfections that if they existed an inspector would have noticed them, even if not specifically looking for them. If this approach gives inadequate strength, advantage can be taken of the difference between design and assessment by measuring the actual imperfections and using those in the model. Strictly, the deformed geometry should only be measured with the dead load applied which is not the same as the unstressed geometry which is input to the computer. This is conservative and not usually a problem. However, it can happen that even very small assumed initial imperfections give larger than observed deformation. This implies that the model is conservative, probably by underestimating restraint.

When selecting the Eigen value buckling modes to decide assumed imperfections, it is important to appreciate that they are not necessarily the lowest ones. For example, U-frame bridges that actually fail by lateral torsional buckling action may have several lower web buckling modes. The non-linear analysis will correctly predict post-buckling behaviour in the webs as they begin to work by tension field effect. However, there is no equivalent for the global lateral torsional mode and, due to yielding, the bridge will fail before it reaches the elastic buckling load. In such a case the imperfections should be applied for the global mode even though the corresponding elastic buckling mode may be much higher than for the first web mode.

If the non-linear analysis achieves the required load without the material exceeding design yield stress anywhere, it has proved the structure is adequate. However, if there is any yielding, it is necessary to consider the implications. Usually the simplest solution is to switch to a model which has material as well as geometric non-linearity. It may be easiest to reduce run times to change the properties only in the critical areas. This should not be made too local as once material yields locally, stresses in surrounding material will increase.

Rigorous analysis often predicts local yielding such as of stiffeners over supports. It is sometimes possible to conclude that this is artificial as, for example, when it is due to supports which have a finite size being modelled as points. It is, however, preferable to model yielding properly, modelling the finite width of supports if needed. This avoids the possibility of, for example, support stiffener yielding being dismissed as local and unimportant when it actually results in loss of bearing stiffener restraint to the girder and premature failure.

### 7.7.4      Joints and restraints

Ideally, non-linear models would include accurate representations of all connections. However, this is rarely practical and can lead to two problems:

1   If the joints are not rigid it can result in over-estimating buckling loads. This can be avoided by using conservatively assessed elastic springs. BD 56/96 gives some values that can be used for connections with cross-girders in U-frame bridges.

2   Where the actual joints are weaker than the material either side, it is necessary to check that the joints do not fail prematurely.

Usually, non-linear analysis is done because a conventional analysis has failed to show adequate strength. This means the details will have been checked and do not need to be repeated. However, the distribution of stress given by the non-linear analysis can be substantially different. For example, analyses that depend on web panels working by tension field action, can give considerably higher forces in rivets surrounding the critical panel.

A further issue related to connections is that, even if rivets or bolts in tension splices are able to resist the full load, the presence of holes weakens the flange. This is often allowed for by using a reduced yield strength so that the program automatically avoids overstress by yielding. If the yielding is very localised (and particularly if it does not coincide with the splice) this is conservative. However, if the flange in the analysis yields over a significant area, care is needed when interpreting the results. In reality, if significant redistribution from the flange is needed to achieve the assessed capacity, the very localised nature of the real yield which is concentrated at the holes can mean the behaviour is less ductile than the computer model implies.

Many older bridges rest directly on masonry or concrete abutments or piers. In such cases it is common to model the support as elastic springs spread over a reasonable area. The strength is not normally sensitive to the stiffness within a realistic range. However, it will often be found that the springs go into tension. It would be possible to release the offending springs by hand but, since this is likely to result in more going into tension, it is convenient to use non-linear gap elements.

It is often beneficial to include the in-plane restraint from the deck which has the effect of keeping girders straight in plan at a particular level. If this level is the same as the bearing level, this can be modelled with rigid lateral restraint. However, if it is not, this would introduce an artificial restraint. It is necessary to use constraint equations, or equivalent, which keep the line straight without constraining the structure in any other way.

It might be possible to model restraint from such things as U-frame action restraining a single girder model by putting elastic springs in the model. Determining the correct stiffness can be difficult particularly if there is a possibility of interacting with other girders which are also buckling. One possible alternative approach is to model the cross-girders to the centreline, and restraining them in rotation but leaving them free to move vertically. This invokes symmetry to model the structure correctly if the opposite girder buckled at the same time. For other cases, such as when the other girder is less heavily loaded, it is conservative. Another possibility is to model more of the bridge, perhaps using a much coarser mesh for other parts.

### 7.7.5 Material properties

Apart from the considerations of partial safety factors discussed in Section 7.7.2, the choice of material properties is usually fairly straightforward. For very slender structures, simple elastic properties may be adequate although this requires yield stress to be checked. For other cases, elastic-plastic properties are often used with steel linear-elastic with a normal code E value up to yield, then straining plastically at constant stress.

In reality, steels do not have infinite ductility. It is possible to put a limit on plastic strain in the computer model. However, the real limit of ductility is not so easily defined as it is gauge-length dependent. This arises because the actual yield takes the form of local necking with the actual limiting strain being fixed more in actual yield distance than percentage. It is often convenient to merely read the maximum strain from the computer and ascertain whether it is excessive. The failure strains from tests can be used provided the yielding is localised. However, if large areas yield, it is necessary to be more conservative.

In reality, steels have significant strain hardening with ultimate strength of mild steel being typically 70 per cent above yield, and higher grades 40 per cent above. Non-linear analyses which do not consider this will never match capacities predicted by plastic analysis as theoretically these require infinite curvature. However, it is not straightforward when including strain hardening safely in analysis as the post-yielding behaviour is sensitive to gauge length type effects. It is usually preferable to use only conservative strain hardening in material models used in the analysis.

## 7.8 CAST IRON

There are still a large number of structures in service which have cast iron components. The great majority are simple infill joist bridges but there are also arches and more complicated structures such as the Coalport Bridge described in Appendix A2.1.

In terms of analysis and assessment, the characteristic of cast iron that makes it different from steel and wrought iron, and indeed reinforced and prestressed concrete, is that it is a brittle material. It has a tensile stress-strain curve as represented in Figure 2.4 and fails abruptly at such low stresses that it is unusual for its compressive strength to be relevant in assessment of beams. It is a very variable material with actual strength varying not only between date of production, manufacturer and element but also according to position in an individual casting. This means that, although the permissible stresses given in BD 21/01 are probably a little conservative for most cases, it is rarely worth taking test specimens to determine higher more accurate properties because the tests tell very little about the strength except of the particular specimen taken. Assessors and inspectors should, however, be aware that there have been cases of fillers with almost no strength being used to hide faults in castings and it may be necessary to take account of this in assessment.

The fact that cast iron is brittle means that cast iron structures are normally analysed using linear-elastic analysis, although occasionally geometric non-linearity may be considered for slender structures. There may be occasions when non-linear analysis could be useful to analyse a structure with cast iron beams and concrete infill, as discussed in Section 7.9.2. The only non-linearity considered then would be that due to cracking of concrete, sliding at supports and possibly yielding of transverse reinforcement.

As well as using linear structural analysis, elastic section analysis is always used. In sections which are not restrained, as by concrete infill, it may be necessary to consider lateral torsional buckling of beams or buckling of struts. An indication of whether this is likely to be relevant can be obtained by adapting steel rules, remembering to correct for the reduced E value as well as stresses. In doing this, it is necessary to be aware that the tensile strength is much lower than the compressive strength.

If the effect of slenderness in struts is significant, empirical approaches have been developed and are implemented in BD 21/01. It is also possible to analyse buckling more directly. A geometrically non-linear analysis, still using linear-elastic material properties, can be used directly to check the stress limitations from BD 21. This is valid because the brittle nature of the material means there is little reduction in stiffness before the stress limitation is reached. This is why use of steel rules to establish if buckling needs to be considered is conservative. However, in doing this analysis, it is important to include geometric imperfections as discussed in Section 7.7.3. It is also important to realise that the low tensile strength of cast iron means that even compression members can fail from tensile stresses.

Although bridge engineers are familiar with using linear-elastic analysis, they are probably less aware that its use at the ultimate limit state is fundamentally underpinned by the safe theorem of plastic design. This says that for a ductile structure if a stress state can be found that does not violate the yield criterion, the structure will not collapse. Linear-elastic analysis finds such a stress state. This means that any geometrically correct elastic analysis will give safe answers whatever element stiffnesses are used. This does not apply to brittle structures: the analysis is only reliably safe if it is correct. This has two important implications:

1    Underestimating the distribution capacity of a bridge, and even use of a static load distribution, does not necessarily give safe answers. It will if all the beams are identical, since the distribution capability will transfer load to less heavily loaded beams. However, if the beams are different, this may not apply. A particular problem can arise in bridges which have a mixture of cast iron and wrought iron or steel beams, which is common in bridges which have had major strengthening

or reconstruction. Consider, for example, a bridge in which the interior beams have been replaced with steel but the original cast iron edge beams kept for appearance. A static load distribution might suggest the edge beams are not loaded by vehicle loading at all. However, in reality they are. The steel beams might typically need four, five or even 10 times the deflection of the cast iron edge girders to reach their peak resistance. This means that transverse load distribution will tend to distribute load to the edge girders even when they are close to failure and the internal girders have a large reserve of strength. Particular care is needed in the analysis of this type of structure.

2   The conventional assumption that imposed deformations, such as differential temperature or support settlement, do not affect ultimate strength is not necessarily true. This is less likely to be a significant issue in girder bridges although it could be in bridges with continuous cast iron girders. However, it can be in other types of bridges such as trusses. In Coalport, for example, the analysis suggested that foundation movement alone could have fully stressed the iron in places. As it is not usually possible to determine the extent of such settlement that has arisen, it is difficult to allow for this correctly but it should at least be considered to ensure grossly unsafe assumptions are not used.

BD 21/01 also gives the D/d approach developed by Chettoe *et al*, (1944). In this the assessment of cast iron girder structures infilled with well consolidated material other than pure sand or pure clay, can be made less conservative by increasing the section modulus used for live load only, by multiplying it by D/d where D is the overall depth of deck less 75 mm for surfacing and d is the depth of the girder. This is a purely empirical approach and it appears to make some allowance for real distribution behaviour as well as composite action. It should not be used in combination with composite action calculations or realistic distribution analysis as this may mean effectively double-counting an effect.

## 7.9   COMPOSITE ACTION

### 7.9.1   Principles

Infill joist and cased-beam bridges, which are now considered as being composite, have been built since well before steel replaced iron in bridge construction. However the beam-and-slab form with stud or other shear connectors, although used as early as the mid-1930s (at least in Tasmania), did not become widespread until after the Second World War. Bridges built before then were not normally designed as acting compositely, and as late as the 1950s, bridges of steel and concrete were built which were not designed to act compositely. This does not stop them acting compositely in fact, nor should it stop assessors considering composite action in assessment, where appropriate. However, it does mean that if proving composite action is difficult, it is worth checking to see if the structure has adequate strength without invoking it. Very often, considering the concrete as merely acting to prevent lateral torsional or local buckling in the metal, will be sufficient to prove adequate strength. It also means, if the drawings give no indication of the connector details, it may be because there are none.

If composite action is considered, the usual approach in UK codes is to determine interface forces from the elastic formula. This requires a specific distribution of shear connectors along the interface: it checks the interface locally for the force predicted from the shear force at that section, rather than checking that the total connection available is adequate to transmit the total required force from the metal section to the concrete slab. Provided the shear connection is adequately ductile (which mechanical shear connectors such as studs normally are) the restriction to requiring the connection to be in a specific place is not really needed. If conventional assessment shows the shear connection to be inadequate locally, alternative assessments are possible.

Where the slab is in compression, merely redistributing the total force from the elastic calculation along the girder may not be safe. Although the elastic approach has been found by comparison with tests to be safe, and it has a conservative distribution, the total force it derives (which assumes an uncracked elastic distribution of flexural stresses) may actually be less than required for a plastic analysis. That is, it may be inconsistent with the flexural analysis of compact sections. The actual force required in the flange at the critical section in flexure from the plastic analysis, can be determined and distributed among the shear connectors between there and the point of contraflexure.

This is less likely to be a problem where the concrete section is in tension, as in typical hogging sections in continuous girders. This is because here the uncracked elastic section analysis overestimates the actual force required to be transmitted by the shear connectors to the concrete slab. Because of this, if elastic analysis suggest the connection is inadequate, it is even more likely for such cases that the alternative approach will give a higher assessed capacity.

If the shear connection still appears inadequate after realistic assessment but there is a reserve of flexural capacity, partial-interaction analysis could be used. In this approach, instead of checking the interface for the force required by the ultimate moment capacity analysis, the flexural section analysis is undertaken assuming the force in the concrete is limited by the capacity of the shear connection. This is actually quite simple to do, particularly when plastic section analysis can be used for compact sections.

## 7.9.2     Infill joists and cased beams

Many bridges exist which consist of iron or steel beams infilled with *in situ* concrete. Except sometimes for transverse spanning decks in U-frame bridges, particularly in railway structures, this form of construction is now rarely used. As a result, it is not closely covered in modern design codes. BD 61/96 and BA 61/96 (HA, 1996) address some of the issues, however, a number of problems with the documents have been encountered. These issues have been studied and solutions presented in a Network Rail Current Information Sheet 23 (Network Rail, 2000). This primarily gives the intended interpretation of clauses (ie it is advice on how to use the document rather than giving alternative approaches) and so could in principle be used for all bridges of this type, not just those owned by Network Rail*. When correctly interpreted, BA 61/96 enables bridges having infill joists to be assessed quite well, including considering partial-interaction theory in some cases. However, semi-empirical restrictions are needed and the full approach described at the end of Section 7.9.1 cannot always be used. This is because the chemical bond of the interface is not as ductile as stud shear connectors.

It is actually rather irregular that the assessment approach for many types of structure is given in BA 61/96, rather than BD 61/96. For this reason, it appears, unusually for a BA, that the document should be called up in formal Approval in Principle documents where these are required.

BA 61/96 also gives analytical approaches for bridges of this type which have inadequate transverse reinforcement for conventional analyses of the type that would

---

**Note**    * During the drafting of this guidance, it was anticipated that BD 61/96 would be revised. However, this has not happened. This is primarily because, although this form of construction is common on bridges owned by Network Rail and, to a lesser extent, local authorities, it is not common on the trunk roads and motorways owned by the Highways Agency.

be used for modern designs. These can be useful, however, tests show that actual distribution can be better than even these methods consider. It is possible to use non-linear analysis that predicts this behaviour and which results, at least in part, from the frictional transverse restraint to the girders (see Jackson, 1996).

Some of the same interface issues arise with cased beams as they do not often have shear connectors. BD 61/96 and BA 61/96 also give advice on these, but again there are some problems and it may be useful to refer to the Network Rail Current Information Sheet 23 (Network Rail, 2000).

## 7.10 FATIGUE

The above assumes that static ultimate strength is critical. However, fatigue is an ultimate condition and is relevant to safety so it is arguable that it should be checked in assessment. The major problem with this is that, for structures that have been in service for any length of time, realistic results will only be obtained if a realistic assessment of the past history is made. This is difficult and for this reason, fatigue assessments are normally avoided. They are not covered by BD 21/96 and are only done on highway structures if there is reason to suspect inadequate fatigue life and in some structures known to be fatigue-sensitive.

Railway underbridges suffer more frequently from fatigue problems and accordingly the Network Rail Underbridge Code (Network Rail, 2004b) and the London Underground Standard (London Underground, 2006) give more advice on the subject which may be useful for other types of structures. It may also be more feasible to obtain a reasonable approximation of the past fatigue loading history of railway bridges although the Network Rail documents give little guidance on how to do this.

Modern fatigue standards concentrate on welded structures and coverage of riveted structures can be inadequate or over-conservative. It may be possible to make direct use of more recent research such as by Xie *et al* (2001).

Orthotropic plate decks are among the most fatigue sensitive of highway bridge structures, and of structures that are not considered in the Network Rail code. A number have failed having cracked quite seriously and required expensive repairs. However, these failures have not led to collapse or had any immediate effect on safety. The structures are damage-tolerant. This, and the relatively imprecise nature of fatigue assessments mean it is better to invest money on a reasonably comprehensive inspection for fatigue cracks, rather than on a full fatigue assessment. Fatigue assessment can, however, help by targeting the inspection on areas where cracks are most likely.

Where a full fatigue assessment is undertaken the approach is to use the design rules in BS 5400-10 (BSI, 1980). If the predicted fatigue life is short, it will be necessary to use the full spectral analysis rather than the simplified approaches which are usually significantly more conservative. In design, this normally means applying the fatigue vehicle to the existing computer model used to check other aspect of design. However, if the model has been modified or set up to allow for redistribution this may not be valid. A typical example of this is the use of pin-jointed analysis in a truss structure that is actually welded. The local moments that a fixed jointed model gives cannot cause a static failure but they can cause fatigue failures. It may be necessary to use a different computer model for the fatigue assessment.

If assessment predicts inadequate fatigue life, strengthening may be needed. However, in considering the management of the structure, it is important to distinguish damage tolerant details from the genuinely dangerous ones that could result in failures. For the former, monitoring by targeted inspections may be all that is required. For the latter, measurements of local stresses under normal traffic may be required to confirm the situation.

## 7.11 Implications for approval and procurement procedures

Assessment should normally progress from simplistic to more advanced and usually more expensive approaches (see previous sections and Figure 7.1). However, neither the formal approval procedures nor the lump sum procurement approaches used by many bridge owners for work done by outside consultants are well suited to this. Like codes of practice, they were originally developed for design. Here the analytical approach and the details required for formal Approval in Principle documents, and the amount of work required for the lump sum, can be known in advance. The structure is then designed to suit the analytical results.

Conversely, in assessment the analysis should be refined to suit the structure. In some cases, the Approval in Principle problem can be avoided by completing the document, not when shown in Figure 7.1, but only after the final assessment approach has been decided. However, it could be argued this defeats some of the object of the system and it also makes working to lump sums even more difficult. An alternative is to prepare the Approval in Principle for, and limit the lump sum to, a specific form of analysis. When the results are reviewed the possibility of further refinements is considered. If it is decided to proceed to further refinement, the Approval in Principle is amended and either a new lump sum is agreed or the work switches to a time basis. In some cases, particularly if a large number of bridges of similar basic form are being assessed, it may be possible to agree such provisional lump sums in advance.

With a good relationship between client and consultant, such approaches can result in satisfactory cost effective assessments. Without it, as with slavish insistence on completing assessment to lump sums obtained from a lowest priced tender, what should be initial simple conservative assessment results are used as final results and bridges are strengthened unnecessarily, which may bring the whole process into disrepute (Jackson, 2001).

It is always possible that some of the refinements give no benefit. They may increase assessed strength but not by enough to increase the live load rating by a whole class. This means some work has been done with no benefit. If this is unacceptable to clients, one possibility would be to pay for special assessment work using advanced methods on the equivalent of the lawyers' *no win, no fee* approach, based on a percentage of the savings.

# 8 Remedial works – general principles

> An overview of the general principles of remedial works involving repair and strengthening is given:
>
> • checklists are included to assist the engineer in identifying critical issues
>
> • strategies for repair or strengthening are introduced, and reference made where further detail is provided in Chapters 9 and 10.

## 8.1 INTRODUCTION

The physical condition and weakened areas of a structure should be determined by inspection and analysis. In the course of this work the engineer, acting on behalf of the bridge owner, has to exercise judgement regarding the form and extent of any remedial measures. The context of the work is very important, and the engineer will be informed in his decisions by data from the inspection reports, the assessment findings, and information taken from the bridge management system. The latter will identify the importance of the work compared to the condition of the bridge stock as a whole.

The client or bridge owner should understand that interventions need to be viewed in the overall context of their strategy for the asset base. Aspirations (see Chapter 4) may be to:

• improve the condition of the bridge stock

• maintain the condition of the bridge stock at current levels

• manage the condition of the bridge stock (or a structure) towards replacement.

These should be reflected in the decisions related to a particular structure.

## 8.2 ISSUES

### 8.2.1 Remedial works – repair, strengthening, protective treatment

General issues related to remedial works involving repair and strengthening are considered in this chapter. More detailed guidance on repair methods is given in Chapter 9 and methods for strengthening and replacement of structures are discussed in Chapter 10. Repairs related to protective treatment are dealt with in Chapter 11. Any repair or strengthening works to the structure will almost certainly have an effect on the protective treatment in the area affected by the work.

Contributing factors to the need for repair or strengthening may be poor water management, poor maintenance, or the efficacy of protective systems. Repair and strengthening measures should always include:

• a thorough review of the existing regime of routine maintenance

• consideration, recommendations, and funding for future routine maintenance to ensure continued integrity of the structure after the works have been completed.

## 8.2.2      Guidance in considering the need for repair or strengthening

Any repair or strengthening operation will be an intervention, and is likely to change the way that a structure performs. Unless properly planned and executed, any intervention has the potential to adversely affect a structure. Remedial works should proceed only when decisions arising from consideration of the bridge condition and maintenance strategies deem intervention to be justified on the basis of safety, structural integrity or economics. Other avenues including reduction of load effects, or even a do nothing and monitor option should also be explored. Key questions in considering remedial works include:

1   Is the structure showing signs of distress?

2   Is the structure fit for its purpose, from the point of view of carrying the required weight and volume of traffic safely?

3   Does the structure have adequate strength and stiffness?

4   Is this judgement supported by a structural analysis of appropriate depth and rigour?

5   Are there any inherent weaknesses in the structure?

6   Is the structure subject to statutory controls? Can loading be restricted without affecting operational requirements?

7   Is the structure listed or scheduled, which influences the type of intervention possible (see Section 3.3)?

8   Is the overall strategy to repair, strengthen or eventually reconstruct?

9   Is there an engineering team already familiar with the structure, the structural analyses, and the repair work that may be necessary?

## 8.2.3      Operational constraints

In addition to the above, operational constraints at the site will influence the design and should be considered. Examples are:

- disruption to the travelling public and whether possessions or partial or full closure will be required, and the lead time for booking them

- disruption to local residents and whether noise and dust restrictions influence the selection of materials, methods and working practices eg use of air wrenches for tightening HSFG bolts

- safety, security and welfare of the workforce including safe access for inspection, access for the works, risk of theft or damage to equipment

- environmental issues (see Section 3.4 and 3.5)

- availability of a workforce skilled in the relevant trades.

## 8.2.4      Options

A list of options accompanied by an associated risk schedule for the proposed measures should be prepared at the earliest possible opportunity. A structured risk-based approach to the work will identify major areas of concern. From this, testing or investigation can be undertaken to collect missing information and mitigate the risks. Where it is not possible to carry out investigations, and there is a likelihood that further damage may be discovered during the works, this should be reflected in the scope of the project at the tender stage, with adequate sharing and allocation of the risks. Risk allocation and acceptance should be resolved before work starts on-site.

Sustainability issues can be taken into account by the use of embodied energy considerations to inform the decision to repair, strengthen or replace the structure. Guidance on use of embodied energy and sustainable construction is given in *Achieving sustainable construction* (CORUS, 2003).

The options list should include allowance for annualised expenditure for subsequent aftercare of the structure.

The form of any tender process to select contractors should be carefully considered, as some repair and strengthening methods are specialised, and there is benefit in bringing the contractor's expertise and knowledge of equipment into the design process at an early stage.

### 8.2.5      Rapid response or leave and monitor

In some circumstances, the results of inspection, assessment, or observation by the public can lead to the need for special measures or an emergency response.

Key issues in these circumstances are whether the inadequacy in the structure is:

- theoretical – the structure appears to be unsafe because its strength cannot be justified by calculation

- observed – the structure appears to be unsafe because of its appearance as a whole, or of one component

- inadequate at the serviceability limit state – there is a loss of use of the structure, and damage to elements which is not critical in the short-term

- dangerous – portions of the structure have become loose or detached

- inadequate at the ultimate limit state – there is a danger of collapse or catastrophe.

Hurried intervention and repair of a structure has the potential to be ill-considered and make matters worse. While temporary measures may be required, a cautious approach should still be adopted to long-term repair or strengthening action which should follow the above guidelines. The principles of such an approach are set out in BD 79/06. In addition, the publication *Steel bridge strengthening* gives examples of application (Kennedy-Reid *et al*, 2001).

A process of risk assessment and its management should be carried out. If the reasons for the failure are not obvious, monitoring of the structure should be considered to collect information about actual behaviour and, where possible, to calibrate the analyses that have been undertaken. Proposals for monitoring should be considered and set out in a reasoned manner, using a format similar to the familiar Approval in Principle documents. Further discussion and methods of monitoring are included in Chapter 6.

## 8.3      REPAIR OR STRENGTHEN

It is useful to distinguish between repair and strengthening, although the process of implementation may, in some ways, be similar.

Repair implies restoration of the structure or element up to, but not exceeding, its original strength, and reinstatement with the same or compatible material.

Strengthening implies a degree of upgrading to a higher strength, because of either higher load-carrying requirements or past deficiency. There will be a need for calculations to justify the work.

In addition to replacement material, it is likely that extra sound material will be required to enhance the strength of parts of the structure. Strengthening may relate to work, on the structure as a whole, or on individual components.

Repair and strengthening operations need to be viewed in the context of the structure as a whole:

1  Will the repair or strengthening scheme affect the overall behaviour of the structure?

2  If so, to what extent?

3  Has so much work been done on the structure that the overall behaviour has been affected?

In repair or strengthening, the impact of the work on the appearance of the structure should be considered. If necessary, adjustments should be made to the design to ensure that the end result is visually acceptable.

Heritage structures under statutory control provide a particular challenge in this regard. The general requirements and challenges of heritage structures are outlined in Section 8.5.

## 8.4    STRATEGIES FOR REPAIR OR STRENGTHENING

Remedial work to a structure should use sound structural principles, and should be based on a thorough understanding of the load paths within the structure. This applies to both the existing load paths and the new ones after addition of new material.

It is important to recognise that the addition of material to a structure will affect the relative stiffness of components which may be sufficient to change the distribution of load. Care should be exercised, for instance, if a computer analysis that is used for assessment is subsequently used for assessing the performance of the strengthened structure. Changes in member properties should be properly represented, and the resulting output of load effects should be carefully reviewed.

When material is added for repair or strengthening, it is important to ensure that it will act compositely with the structure, and that load effects are transferred to and shared with the new material. Methods such as jacking or some form of prestressing may be required.

Remedial works often require considerable ingenuity in design or implementation. It is important to also recognise that a degree of robustness is almost always required, and so innovative methods and materials should be thoroughly explored before incorporation into the works.

Remedial work to an element or structure can be considered under three strategies:

1  Reduce the perceived load effects.

2  Increase the resistance.

3  Replace the element or structure.

The strategy for repair or strengthening may be a combination of one or more of the three. Methods for repair (as previously defined) are discussed in Chapter 9. Methods for strengthening and replacement are discussed in Chapter 10. Some of the methods for strengthening may also be used for repairs and vice versa.

## 8.4.1     Reduction of perceived load effects on the affected element

This strategy relies upon either reducing the loads applied to a structure, or adjusting the relative stiffness of different parts of the structure such that load is diverted away from critical elements. Key features of this approach are:

### *Review assumptions made during the assessment process*

1    How reliable are the assumptions made for the particular situation?

2    Are the applied loads derived from national (or other) standards?

3    If so, are they appropriate for the bridge location?

4    Are the partial load factors appropriate for the particular situation?

5    Is it possible to calibrate the calculated load effects against actual measurements?

### *Better structural analysis*

In appraisal of a structure, it is frequently possible to demonstrate adequate capacity with a relatively simple analysis. If such analyses suggests that a component is under strength, a detailed re-analysis should be carried out using more sophisticated methods, as explained in Chapter 7.

### *Better load distribution through improved or additional load paths*

It may be possible to demonstrate improved load paths through analysis or *in situ* load testing. In addition, calibration of any analysis by reference to observation or load testing can be extremely useful in understanding the behaviour of the structure, and will lead to increased confidence in deciding whether strengthening measures are required. It should be noted that there can be both beneficial and adverse effects, as any material added to the structure will increase the relative stiffness of that part of the structure. As stiffness attracts loads, there will be some increase in the calculated load effects.

### *Restriction on use*

In some circumstances, it may be acceptable to lower the weight limit on the bridge as an alternative to expensive or disruptive repair, or strengthening works. In particular, abnormal loading requirements should be carefully considered and it is necessary to determine whether the load combination is appropriate for the rare, extreme loading conditions that will actually occur on the structure.

In Chapter 4, Box 4.3 for Boston Manor Viaduct describes how unwanted restraint in the structure was contributing to deterioration in half-joints supporting a 36.5 m suspended span. To understand the structural behaviour, and to calibrate computer analyses, load testing was carried out during a motorway possession using vehicles of known axle weights and travelling down each of the four traffic lanes. This proved useful as it demonstrated significantly more transverse load distribution in practice than predicted by analysis.

At each of the four upper and four lower half-joint positions, plating was added to both sides of each of the truss nodes to provide an alternative load path around a damaged element, see Matthews and Ogle (1996a and 1996b).

## 8.4.2 Increase in the resistance of the affected element

Consideration of the resistance of an apparently under strength element can be approached in several different ways:

### Review design assumptions

Design guidance and codes of practice are written to convey best practice within the industry, and aim to anticipate most, if not all, situations that the designer is likely to encounter. Most codes of practice are targeted at new design, where the flexibility to add material or adjust the form of the structure usually exists. The formulae within codes of practice for new design use assumed levels of initial imperfections that are based upon realistic but achievable magnitudes of construction tolerances.

In using the same guidance for re-design of a detail for a repair or strengthening scheme, the engineer is faced with structures of a defined form that were perhaps designed before current codes of practice were written, and built to tolerances that may not be within current accepted limits.

These differences can sometimes be used to advantage. In a number of cases more detailed examination of design assumptions has led to the engineer being able to demonstrate adequate resistance, and avoid the need for remedial work.

Typical considerations are:

1   How reliable are the design assumptions for the particular situation?

2   Are the partial material factors in the design codes appropriate for the particular situation?

3   Are the levels of imperfections measured in the structure smaller than those assumed in design, giving scope for higher capacity?

### More rigorous analysis of the element strength

Having reviewed the criteria against which the structure is being assessed, a more detailed consideration may be made of the way in which the strength of an individual element is derived. Again, the emphasis is on examination of the actual structure as it exists, rather than general assumptions inherent in design literature. Typical features are:

1   Can the structure and its constituent elements be reassessed, or re-analysed using more sophisticated methods? For instance, can a general analysis that is adequate for the structure as a whole be supported by (say) a more detailed localised finite element analysis of elements shown to be critical?

2   Is it possible to identify additional strength by reference to actual material test certificates from the original construction, or from tests on material from the structure?

3   What is the effect on the calculations of capacity if imperfections are measured and found to be smaller than those originally assumed?

### Provision of additional material

After investigation, the engineer may be satisfied that it is necessary to carry out work on the structure by adding material to increase strength. Typical strategies for doing this are:

- reduce stress levels by adding material to increase the capacity of the element in yielding failure. Adding extra material to a component will reduce the stress for a given level of applied load. Note, however, that adding large amounts of material will add extra stiffness, attract extra loading to the component, and may require additional material to compensate. The balance has to be investigated carefully

- increase the buckling strength of a component that is slender by adding material. In the limit, a slender component can be restrained to such an extent that buckling is not a problem, and yielding failure governs. Buckling restraint can be improved by adding plating to the member, bracing critical points on the member, or by encasing the member in concrete

- change the shape of a section, and altering the load path through it. This can be achieved by addition of plating to the element, providing local thickening or bracing.

Further details and means of connection of new materials are discussed in Chapter 10. In all cases, when material is added, consideration should be given to the manner in which load is shared between existing and new material, and how loads are transferred to and from the new material by the connections. Figure 10.8 illustrates this process for a cover plate on a beam.

## 8.4.3 Complete replacement of the element or structure

If a structure shows significant deterioration, then partial or complete replacement should be considered. Thorough investigation is important, and the project should be approached with a good knowledge of what may be encountered, including allowance for risk, both in terms of time and cost.

When elements of a structure are replaced, the manner in which load or stress is taken up requires consideration. In some cases jacking or post-tensioning procedures may be necessary to ensure that the new component participates fully in structural action.

## 8.5 HERITAGE ISSUES

Work carried out on listed and scheduled bridges is subject to the requirements of legislation as given in BD 89/03 (HA, 2003).

The principles of conservation are:

- there should be minimal introduction of new material

- there should be minimal changes to the structure or its appearance

- bridges are best kept in use and maintained in their original form and performing the same function and subject to the same structural actions

- all work should be undertaken using appropriate materials and methods of application

- modifications should involve no loss in character, minimal loss of historic fabric, and minimal adverse effect on the setting

- any modifications should preferably be reversible

- records should be kept of the work and any archaeological artefacts that may be uncovered. There may be a requirement to carry out an archaeological investigation prior to the work

- a conservation plan should be drawn up defining the significance of the bridge and policies to retain that significance. Further guidance is given in Chapter 3 and the English Heritage publication *Informed conservation* (Clark, 2001) and *Conservation of bridges* (Tilly, 2002).

## 8.6    FACTORS IN THE EXECUTION OF REMEDIAL WORKS

Key factors in the execution of repair or strengthening for remedial works are:

- safety of all site workers, visitors and the general public

- likelihood of any defined weather windows when work is likely to be disrupted or impossible

- availability of suitable access (including closures or possessions)

- availability of materials, possibility of any key components on long lead times for delivery

- size and weight of components, impact on delivery and installation methods, craneage

- curing times for concrete and chemical bonding agents

- properly derived probability of completion within the required programme time

- adequate allowances for environmental issues

- planning and co-ordination of tasks, and of skills of the workforce

- use of structured risk management techniques, leading to appropriate risk sharing, ownership, and contingency plans.

## 8.7    PUBLIC RELATIONS

### 8.7.1    Public interface

Regular meetings between contractors and representatives of local residents' associations should be held to ensure that public concerns are addressed. It is common practice to inform residents by letter drop and to provide a telephone number for enquiries. In addition, it may be appropriate to visit residents living close to the bridge works and most affected.

Residents should be kept informed on a weekly basis by regular communication (mail/email/website) of planned activities with noise categorisation for the planned work. This is particularly important if work is at night. In extreme cases it may be appropriate to provide alternative accommodation for residents who are severely affected.

### 8.7.2    Noise considerations

Primary acoustic damping can be achieved by erection of acoustic barriers. Quieter tools and methods can often be substituted, for example:

- if a hammer has to be used to knock out a rivet or a piece of steel, fewer blows with greater force can be employed

- timber blocks can be used to limit the noise of metal impacted against metal

- tools producing percussive noise can sometimes be substituted by non-percussive tools.

### 8.7.3    Traffic

Nightly closures can reduce the impact on local businesses and inconvenience to local residents. This has a downside because it makes the bridgeworks more difficult to undertake as the site has to be set up every night and decommissioned before the start of business the next day. Also, the irritation caused by work at night may outweigh the advantages of day work. As a result, the works can be slower and more expensive to undertake.

# 9        Remedial works – repair

> This chapter provides a review of remedial works related to repair and deals with:
> - factors for selection and execution of repair techniques
> - techniques for repair and component replacement.

## 9.1     MATERIALS USED TO CARRY OUT REPAIRS

Materials for repair works can include steel, concrete (usually through composite action, or restraint to buckling), fibre composite materials and post-tensioning carried out using prestressing strand or high-tensile bars (eg Macalloy).

In addition to material used in the permanent works, repairs frequently involve using a range of temporary works supports, which will include falsework and formwork, timber packing etc.

In some cases, historically sympathetic materials such as cast iron or wrought iron may need to be used on heritage structures as part of the requirements of heritage authorities.

## 9.2     FACTORS IN PLANNING REPAIRS

Key factors in planning repairs to a structure are:

- constraints on cost, time and access for the works
- the extent of work to be carried out
- the uncertainty regarding the extent of the defects which are often not apparent until the work starts
- the extent of regulatory requirements (see Chapter 3)
- diversions for statutory undertakers (see Chapter 3)
- phasing of the works, and logistics of supply, installation and removal of material, including use of large cranes or transporters
- compatibility of the temporary and permanent features of the repair or strengthening scheme
- capability and experience of the contractor
- flexibility permitted within the tendering process to appoint a contractor early in the design process.

## 9.3     TECHNIQUES FOR REPAIR

### 9.3.1     General

There are a number of tried and tested repair methods that can be used to improve the condition of a bridge structure:

- strapping

- repair of fatigue cracking

- stitching

- refurbish corroded components, clean and paint, welded insert, re-design

- refurbish riveted connections, corroded faces, broken rivets

- heat straightening

- improve water management (see Section 4.3).

Methods of connecting repair materials or components are discussed in Chapter 10 as they also apply to strengthening. The repair method should be selected to meet the specific requirements of the bridge and application in question.

### 9.3.2 Strapping

A combination of localised loading including thermal effects, fabrication flaws, or impact damage may lead to cracking in bridge components. Often the load carrying capacity of the support can still be mobilised, providing local repair is undertaken. Strapping of some form is a method that is often used to provide such a repair, or may be used to strengthen an undamaged member. Strapping is commonly carried out using steel bands, as shown in Figure 9.1 (see also Figure 5.4). This has the advantage that some access to the cracked member to monitor behaviour is still possible. Other forms of strapping have used a combination of concrete encasement and circumferential or diagonal post-tensioning.

**Figure 9.1**

*Strapping can be provided in a number of ways, frequently by steel bands as shown here*

### 9.3.3 Repair of fatigue cracking

Inspection of a structure will sometimes identify cracking which, in terms of safety, may or may not be significant, but should be of concern until the causes are identified. Initially a check on the toughness of the material should be carried out to ensure that it is not brittle at the low temperatures which can be experienced by the structure. The combination of a crack initiation site, brittle material and loading or internal residual (built in) stresses can result in a brittle fracture, sometimes called fast fracture, as described in Chapter 2.

Structures or components of structures that are subject to regularly repeated loadings may be subject to fatigue cracking. A crack will initiate at a discontinuity, and will propagate gradually with each load cycle. Crack initiation sites may not always be visible as they may be hidden by protective coatings, or in the case of cracking initiated at rivet holes, hidden by components of the structure. Before a judgement can be made on repair, these factors need to be investigated. Information about the susceptibility of various details to fatigue damage is give in BS 5400-10 (BSI, 1980) and BS EN 1993-1-9 (BSI, 2005), by reference to a series of classifications and examples. It may be possible to raise the classification of the affected detail through re-design of the structure, or diverting load away from the critical area (see Chapter 10).

In some cases, where calculation or observation suggests that crack growth is likely to be small or non-existent, a monitoring strategy may be appropriate, as hurried repair works may make matters worse.

Ultimately, a repair may have to be carried out by removing the affected material (including that containing the crack tip) and replacing with a better detail, or with higher quality material by controlled welding. It is sometimes possible to halt crack propagation by drilling a hole at the crack tip, but expert advice should be obtained as this technique is not as straightforward as it may appear.

It is necessary to be aware that fatigue cracks can be difficult to identify with complete certainty, and if in doubt for critical situations, specialist advice should be obtained.

## 9.3.4     Stitching

Stitching was developed specifically for repairing cracked or fractured metals. Cold metal stitching has a long track record, having been used for repairing equipment and machinery in the engineering industry for more than 100 years. The crack is stitched with profiled connectors that are secured in formed holes using one of a number of proprietary systems that are in use. The process is an accepted repair technique for cracked, broken and damaged castings in cast iron, aluminium and steel. However, it is not possible to quantify the strength of a stitched repair so it is often necessary to couple it with other works. The stitching process is outlined in Figure 9.2 and described in this section.

The method requires a series of sets of adjacent circular holes to be drilled in-line across the crack. Apertures are jig-drilled across the crack to accept multi-dumbbell shaped keys made from high-nickel steel having the same coefficient of expansion as the host material, often cast iron. High-nickel steel is specifically chosen because it is strong enough to take shear loads, but sufficiently ductile to provide the necessary elasticity. The shaped key connector is driven into the holes across the crack, where it is anchored by its shaped profile. The keys are peened into the apertures to become integral with the parent metal.

When the stitching across the crack has been completed, holes are drilled along the line of the fracture between each stitch. These are tapped to receive special stud screws which fill the crack, each one positioned to overlap its neighbour, and to ensure that it is completely watertight. The combination of keys and studs produces a rigid and pressure-tight repair. Final peening and hand dressing completes the operation. Once the repair has been dressed and a primer has been applied and finish painted, it is practically undetectable.

An advantage of a repair using this technique is that no thermal stresses are introduced, no distortion occurs and, in many instances, machining is unnecessary.

The stitching should extend beyond the end of an internal crack to inhibit future extension of the crack beyond its present limit. Non-destructive testing is essential to determine the extent of the cracks, since these may be more extensive within the metal than visible on the surface.

In cases where whole sections of material are missing due to the effects of corrosion or mechanical damage, a patch of metal, known as an insert, can be made to fill the gap and stitched into position. Inserts of this type can be anything from 50 to 1000 mm in size.

An example of the use of stitching to repair a damaged cast iron beam is given in Appendix A2.2.

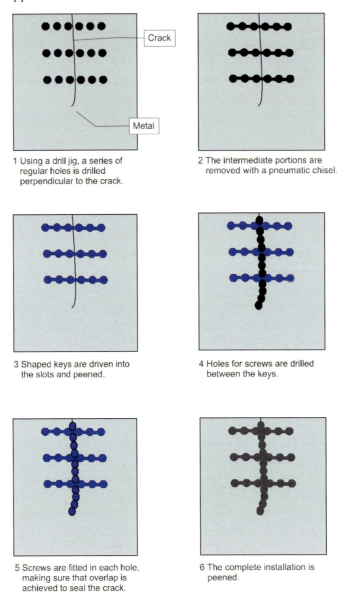

1 Using a drill jig, a series of regular holes is drilled perpendicular to the crack.

2 The intermediate portions are removed with a pneumatic chisel.

3 Shaped keys are driven into the slots and peened.

4 Holes for screws are drilled between the keys.

5 Screws are fitted in each hole, making sure that overlap is achieved to seal the crack.

6 The complete installation is peened.

**Figure 9.2**    *The metal stitching process*

## 9.3.5    Repair impact damage

Many low headroom bridges are liable to suffer damage by the impact of passing vehicles. Often this damage involves distortion, buckling and tearing of the outer girders, generally confined to the bottom flanges and lower areas of web panels. Impact of the sagging (positive bending) zones is less critical than that occurring in the hogging (negative bending) zones since the latter situation can make the distorted flange more liable to buckle under compressive stresses. Depending on the amount and type of damage caused by the impact, it will be necessary to carry out a full dimensional survey to determine the extent and quantity of distortion so that the appropriate repair method can be used. Two methods are outlined below.

### *In situ* repair

The footbridge shown in Figure 9.3 is a Warren Truss of 33 m span, which had been impacted at a nodal point by a lorry-mounted jib. The headroom was not sub-standard, in fact it was in excess of the recommended 5.7 m. The damage comprised deformation of the steel chord as well as several cracks emanating from the point of impact. It was found feasible to repair the damage *in situ*. The bridge was propped at the nodes on either side of the damaged section and jacked up from a position on the opposite carriageway calculated to take out rotations, vertical deflections and stresses due to the damaged section. This method enabled the repaired element to take its share of load.

**Figure 9.3**      *Impact damage to bottom chord of a Warren Truss*

The impact damage shown in Figure 9.4 is to the bottom flange of a Vierendeel Truss that was severely damaged by an over-height vehicle. In this case it was not practical to repair the damage *in situ* and the footbridge was lifted out by crane and transported to a fabrication yard. By laying the structure on wooden railway sleepers, the need to preload the element or manipulate the deflections due to dead load was made unnecessary. The severely torn flange was repaired and the structure returned to service.

**Figure 9.4**      *Impact damage to bottom flange of a Vierendeel Truss*

### Heat straightening

Heat straightening is a suitable method that can be used to repair more extensive buckling and distortion damage in wrought iron and steel. In heat straightening, a limited amount of heat is applied by blow-torch in specific patterns to the deformed regions of damaged plate elements. The heating is applied in repeated cycles of heating and cooling to produce a gradual straightening of the material.

The process imposes internal and external restraints that produce thickening of the plating during the heating phase and in-plane contraction during the cooling phase. Mechanical force is not used as the primary instrument of straightening, which distinguishes it from other methods.

It is also possible to fit additional stiffening elements (for example bracings or web stiffeners) in conjunction with the heat straightening process.

A benefit of using the process is that the bridge does not usually require any temporary supports during the repair. Another benefit is that the process can eliminate or greatly curtail the need for full road closures. When carried out satisfactorily, the heat-straightening technique is an effective and economical approach that can be used in many cases of damage. The same principle can also be applied to the fabrication of new bridges for curving or cambering a girder.

The amount and location of the heating has to be carefully controlled to avoid loss of strength of material. Metallurgical advice should be sought if the repair technique is to be used on steel structures which may have received specific heat treatments during manufacture.

There is considerable experience with the technique in North America, and the Federal Highway Administration (FHWA) has produced a detailed technical guide (US DoT, 1998).

The process of heat straightening as applied to the Brockhampton Bridge, Hampshire, is described in Appendix A2.5.

### 9.3.6      Repair of corroded components

If corrosion is localised, it is often possible to make a localised repair by removal of the corroded section back to sound metal, cleaning the host material around the affected area, and replacing the section using a welded, bolted, or in some cases, a resin-bonded connection. Note that all corrosion products should be removed and fixing of the plating should not depend on adhesion to corroded areas as the process can reactivate and cause debonding.

It should be recognised that, if the material has been seriously corroded, some of the load that it was originally carrying has been shed elsewhere, and there may be a need to analyse the structure to check that other components have not become overstressed.

During repair of corroded components, it may be necessary to use jacking, post-tensioning or other techniques to ensure that the repaired section takes its share of the applied load.

An example of a welded and bolted repair is shown in Figure 9.5. Practical considerations regarding the cleaning of material *in situ*, and on-site repairs and strengthening are discussed in Chapter 10.

**Box 9.1**    *Local repairs to a truss bridge*

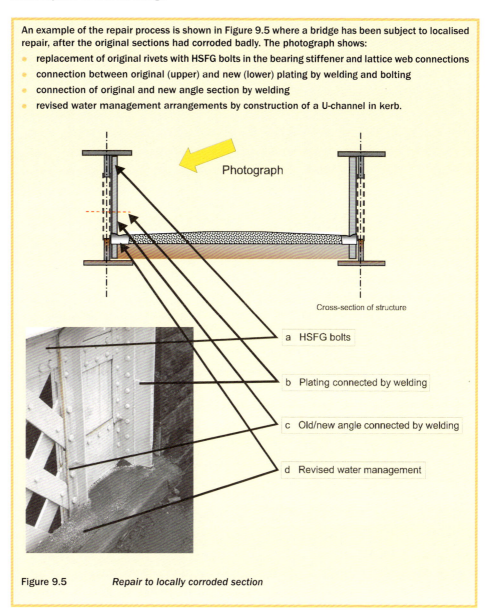

An example of the repair process is shown in Figure 9.5 where a bridge has been subject to localised repair, after the original sections had corroded badly. The photograph shows:

- replacement of original rivets with HSFG bolts in the bearing stiffener and lattice web connections
- connection between original (upper) and new (lower) plating by welding and bolting
- connection of original and new angle section by welding
- revised water management arrangements by construction of a U-channel in kerb.

Photograph

Cross-section of structure

a   HSFG bolts

b   Plating connected by welding

c   Old/new angle connected by welding

d   Revised water management

**Figure 9.5**      *Repair to locally corroded section*

During the repair work, requirements to maintain traffic loading on the bridge may make it necessary to provide temporary strengthening measures to permit removal and replacement of corroded sections. An example of temporary strengthening is shown Figure 9.6.

In some cases, the additional material can be part of the repair, although consideration of the appearance of the repaired structure may not permit this. In the example shown in Figure 9.6 replacement of the flanges and removal of the temporary angles was required.

Note that if the temporary angles had been used as part of the permanent solution (instead of the flanges) they would probably have had to be larger in order to support full scale static and dynamic effects. This method was preferable to the lower load levels that may be justified during refurbishment works by enforcement of loading or speed restrictions.

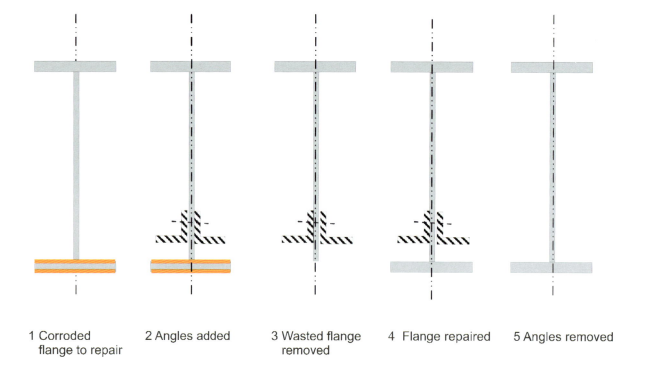

| 1 Corroded flange to repair | 2 Angles added | 3 Wasted flange removed | 4 Flange repaired | 5 Angles removed |

**Figure 9.6** *Temporary strengthening during repair of corroded section*

## 9.3.7 Refurbishment of riveted connections

Figure 9.7 shows typical damage that can be experienced if a riveted connection is allowed to deteriorate. The two pictures were taken on a structure (a former light railway bridge now used as a footbridge) which is formed from a number of riveted and simply supported truss girders. The images show positions where the trusses meet over a common support on successive spans of the same structure, and illustrate the progression of degradation. The connections shown are of the truss end (bearing) stiffeners at the ends of the top flange plates.

In the connection shown in Figure 9.7(a), exposure to water has resulted in the formation of corrosion products between the faces of the plates and stiffener angles. The protective treatment to the edge of the plates and the joint has broken down, water has been drawn into the joint and corrosion has taken place. The corrosion products occupy a larger volume than the original parent metal and the gaps have been forced open by the expansion. The larger gaps admit more water, and the corrosion process has been accelerated. The rivets are clearly vulnerable, and Figure 9.7(b) shows the eventual result where the rivets have fractured under the bursting pressures. It is notable that the actual measured loss of plate section is not large.

In these circumstances, a relatively simple repair involved cleaning the metal, replacing the rivets with HSFG bolts, and repainting to prevent more expensive work in the future.

(a) **Plates forced apart by corrosion products**

(b) **Rivet fractured by bursting pressure**

Figure 9.7          *Deterioration of riveted connections*

Corrosion frequently attacks the heads of rivets, and leads to a reduction in the overall head size. Unless there is obvious deterioration to the joint, the decision to remove rivets should be approached with due care.

When originally formed, the single headed rivet is heated to red heat, and its shank is introduced from one side of the joint. On the other side of the connection, the second head is shaped from the emergent shank tip by hammering or pressing. This process results in some plastic deformation of the rivet into the hole so that it frequently approaches a tight fit. Subsequent cooling of the rivet results in the plates of the joint being drawn tightly together, so the action of the rivet is three-fold:

- shear resistance is provided by the shank due to the tight fit in the hole

- axial compression is provided by the shank of the rivet and anchored by the heads at each end

- additional shear resistance is provided by the clamping action of the rivet on the plates.

When faced with the decision to replace rivets or not, the engineer should understand the load paths through the connection, and use judgment based upon the current condition of the rivet and plates. Assuming that there is no corrosion between the plates of the connection, the following guidance may assist in judging what to do with the rivets.

If the connection is essentially acting in shear and the rivet heads, although corroded, are sufficient to ensure integrity of the connection and prevent water entering into the shank of the rivet, it is likely that the rivet will continue to provide an effective shear connection provided that loose rust is removed and protective treatment restored.

If the connection has any element of direct tension acting on the rivets (such as in a cantilever bracket), the rivet will be acting in a combination of tension and shear. Significant loss of the rivet head may result in loss of anchorage, and a weakening of the connection.

It is often difficult to carry out a repair using new rivets, as nowadays there is limited remaining experience of the technique. Following removal of the existing rivets, new material may be added by re-using the rivet hole with a modern bolt or bolt-like fastener (see below). Note that removal and replacement of active (as opposed to broken) rivets will cause residual load in the original rivet to be transferred elsewhere in the structure.

A staged removal and replacement procedure should be used, particularly where large groups of connectors are involved. Loading on the structure should be minimised during the operation. Prior to removal of rivets, and dismantling a joint, it is prudent to make a series of timber templates to assist re-assembly.

Rivet removal can be a difficult and time consuming process, and will require consideration of proper access and noise control. It should be remembered that during installation the hot rivet will have been inserted from one side of the connection, and hammered from the other side. It is likely that the rivet will have either filled the hole, or at least been forced into an interference fit. Removal of rivets is a process that should only be undertaken by experienced operatives with the correct equipment.

Typical rivet removal involves two steps:

1    The head of the rivet at one end is removed using a pneumatic hammer (sometimes referred to as a rivet buster) and a chisel tip.

2    After the head has been removed, the chisel tip is replaced with a flat punch and the shank pushed out.

It is important to ensure that appropriate precautions are taken to prevent noise, and damage from the expelled portions of the rivet falling from the structure.

Flame cutting has been used to remove rivet heads, but requires skill to avoid damage to the parent metal, and should be only be used when full control is assured.

In some cases it is necessary to remove the rivet by drilling, however this can be time consuming and expensive.

It is likely that the replacement connector will not fill the hole, and so joints should be designed to be non-slip connections, unless there is good justification for them to be otherwise.

Following removal of the rivet and component, holes that are to be re-used should be checked for collateral damage in the form of minor cracks, burn marks, etc. If necessary, local dressing should be carried out before reassembly to ensure that the details will not cause problems in the future. Significant cracking around a rivet hole should be investigated.

Rivets are usually replaced with high strength friction grip (HSFG) bolts, or proprietary fasteners such as Huck bolts, or tension control bolts (TCB).

Preloaded bolts can be used on a one for one basis to replace rivets (note the previous comments concerning the clearance of the bolts in the hole). It is possible to demonstrate the strength of an existing connection in this way, although a well formed, uncorroded rivet may be better than a bolt, and should not be disturbed without good reason.

In situations where appearance is important it is possible to use fasteners that have a dome shaped head, so that the visual appearance of rivets can be maintained on one side of the joint.

## 9.4　COMPONENT REPLACEMENT

### 9.4.1　General

Before embarking on replacement works, it is important to carry out an appropriate amount of investigation.

This is the starting point for the risk management process. Only by acquiring a sound knowledge of the likely risks involved in the replacement work can problems in terms of time and cost of the project be avoided. Time and effort spent in investigation will support decisions on the extent and planning of remedial works.

Whatever intervention is adopted, the following key issues should be considered:

1　**Risk management**

According to Simon *et al* (1997), risk management is a process whereby "responses to risks identified during risk analysis are formulated, justified, planned, initiated, progressed, monitored, measured, reviewed, adjusted and (hopefully) closed". Key steps in the process are:

- understand key client requirements. These will vary eg road, rail, waterway, developer, airport etc

- prepare a risk schedule

- carry out an investigation to mitigate or remove risks

- achieve a level of risk that is acceptable to all

- agree ownership of residual risk. In many cases there is shared ownership with the client

- develop mitigation plans as a fall-back position.

2　**Utilise a form of contract that is appropriate to the work.**

3　**Allow for properly programmed activity:**

- time taken for approvals

- time taken for statutory undertakers diversions

- time required for possessions/closures

- possible disruptions

- fall-back options.

The following sections contain guidance on replacement of components and structures. Further guidance is given in two publications, BCSA (2005) and Iles (2004).

### 9.4.2　Access and removal of protective treatment

Access to the structure, and to the areas to be repaired in particular, will require care and planning to ensure that work can be undertaken safely and efficiently.

In many cases, removal of the existing protective treatment is required to permit repair or strengthening activity prior to application of a new paint system. Shot or grit blasting of the repair area from an enclosed scaffold is often necessary to clean the metal. Additional challenges not least in health and safety are presented by confined spaces inside box girders present. The following should be considered:

- shot or grit blasting can be a noisy process, and can be subject to restriction by local authorities

- old paint is likely to contain lead, and additional controls are required (see Chapter 3)

- enclosure of the scaffold should be set up to ensure that abrasive material is fully contained and the enclosure material is secure against wind damage

- load will be imposed on the structure from the weight of the enclosure, and the operatives and equipment working within it

- load will be imposed upon the structure from build-up of spent shot or grit

- the proximity to other operations

- method statements should be prepared for execution, supervision and inspection of the work

- spillage into watercourses or other environmentally sensitive areas should be avoided.

The issues and problems relating to removal of protective coatings are considered in Chapter 11.

## 9.4.3 Replacement of cross-members

One of the more common structural repairs involves replacement of cross-members between main girders of the deck. Cross-members (or more usually their end connections) commonly experience deterioration due to breakdown of deck waterproofing systems, water ingress and subsequent corrosion.

The perceived difficulty and expense of exposing these members for regular inspection may cause their deterioration to go unnoticed until the repair option has become uneconomic or structurally not possible. This is an important issue as such failures could lead to secondary members becoming wholly detached and falling onto areas used by the general public.

It is important when considering inspection activity to undertake sufficient investigation of the structure to give confidence in the data to be used in aid of decisions on maintenance or repair. If investigation is not possible, appropriate allowance should be made by following the risk management process, and agreeing where the responsibility for any consequential costs will lie.

Whenever a component is replaced, it is important that all the structural actions of the element are identified, and, if necessary, alternatives provided during the repair operations. Even though an element may have deteriorated, it may still be stabilising or sharing load with other parts of the structure. These loads need to be carried elsewhere in cases when the element has to be completely removed. Load effects that are locked into the structure can be significant and need to be carefully considered. Provision should be made to release load in a controlled manner. Examples where this can occur:

- cross-girders and stiffeners of a half-through deck acting as part of U-frame restraint to the compression top flange of the main girders

- end cross-girders and stiffeners of a half-through deck providing overall stability to the main girders at the support positions

- continuous composite members containing in-built stresses due to their original construction sequence (eg wet concrete loads on unpropped construction)

- unintended effects such as beams acting as props between bridge abutments, or arching action.

## 9.4.4     Replacement of bearings

Removal and replacement of structural bearings is one of the more common repair works. More recent structures should have been designed with stiffened jacking points already provided, in such a manner that girders can be jacked off strong points on supports, or temporary trestle positions. If jacking stiffeners have not been provided, it will be necessary to design and install a suitable system for lifting the bridge at the bearing position. This system should be designed so that it can be stored and re-used for subsequent replacements in the future, as the operational life of bearings is likely to be less than that of the bridge.

At pier positions, multi-girder bridge structures are likely to have heavy transverse bracing which is very stiff. Even though the vertical movement to remove a bearing can be very small (1–2 mm in some cases), the loads generated by jacking one girder relative to its neighbours can be significant, and some loosening of the bracing, or simultaneous jacking of more than one girder may be required.

By their very nature, bearing positions are such that large loads and eccentricities are likely to co-exist. The jacking scheme should be carefully prepared, taking account of how these loads and eccentricities are going to be controlled when the bearing is released for removal. In particular, care should be exercised to identify any horizontal restraints.

In some instances, structural bearings were installed using a fine skim of jointing compound to ensure good contact between structure and bearing plates. These materials have some adhesive properties, and methods of breaking the bond and cleaning the contact surfaces may need to be planned as part of the method statement for the replacement works.

Minimising the magnitude of the loading that has to be dealt with during jacking is desirable, and wherever possible jacking under a no live load condition should be considered, as this minimises loading as well as the likelihood of bearing rotation.

In some situations, total closure of a structure is not possible. However it may be possible to undertake jacking at one location (for example on a wide structure) with limited traffic permitted at locations remote from the jacking position where the influence of the loading is negligible. Live loading near to or above the jacking operation is not advisable. Some form of temporary bearing (such as a timber grillage) should be provided if a bridge has to be opened to traffic for an interim period during the replacement works.

Bearing replacement operations should be scheduled to take place when climatic conditions are such that rapid variations in air or bridge temperature are unlikely to occur, as the resulting structural movements can make adjustment and setting of bearing eccentricity difficult. Following replacement of the bearing, transfer of the loading from the jacking system onto the new bearing, without imposing any unintended eccentricity, requires particular care. Options should be well-rehearsed, as the work is usually carried out towards the end of a closure period, when available time for adjustment may be short.

After re-loading, checks should be made for correct seating of the bearing, observing the range of bearing movement by reference to temporary marks or slide paths, and behaviour of the structure over a period of about 24 to 48 hours before the jacking arrangements are dismantled.

## 9.5      IMPROVEMENT OF WATER MANAGEMENT

The need for good water management on a structure has been discussed in Chapter 4, and it follows that systems should be properly maintained and repaired when necessary. If inspections indicate that damage is being caused by poor water management, improvements can include:

- use of larger diameter drain pipes
- avoidance of sharp bends in the pipework
- re-design of the drainage system
- proper provision for regular maintenance of pipework and water traps.

It should be noted that regular cleaning and maintenance is required to ensure effective operation of even the best water management systems.

### Closed members

Moisture in the atmosphere can lead to the presence of water inside closed tubular members, or within larger enclosed structures such as closed hollow sections. In practice moisture-laden air from the atmosphere enters the enclosed space so the water condenses and becomes trapped. There have been instances of tubular members being gradually filled with water in this way and being damaged during cold spells when the trapped water freezes.

It is considered practically impossible to fully seal members, and it is preferable to provide drainage holes at low points in the component or structure. The maintenance strategy for the structure may need to be adjusted to ensure that regular checks and draining of problem members takes place. It is also important to check that the holes do not weaken the structure.

### Problems arising from retrofit activity

Many retrofit activities have to be undertaken during road or rail closures, and are subject to time constraints. In addition, not all contractors working on the structure will have a full appreciation of the long-term impact of their work. An example of this is lighting columns being fit onto a bridge deck over the UK motorway system 20 years ago. They were installed by a lighting contractor in the central reservation of the carriageway. The columns were held down by four large bolts in holes drilled through the concrete deck. The concrete deck was designed to act compositely with steel beams running across the width of the deck, and with plan-bracing on the supporting main structure. A more recent inspection for assessment of the structure revealed that water had penetrated the deck at the bolt locations, and had started to cause corrosion of the plan-bracing and some of the cross-girders. Due to either bad procedure, or hurried implementation, re-sealing of the membrane at the holes had not been carried out. This is a simple exercise that when omitted had the potential to cause significant damage.

It follows that planning retrofitting activities should consider all aspects of the work and the continued performance of the structure. Risk assessments and method statements should make an adequate allowance for time to complete work to the specified standard.

## 9.6      REMEDIAL WORKS TO DECKS

Bridge decks sometimes become inadequate to carry the loads they were designed for, due to deterioration processes such as corrosion or fatigue. Various types of remedial works to decks have been developed to suit particular situations. Although not primarily to raise the load carrying capacity above the design value, these methods can extend beyond repair, and include features that can be described as strengthening. Specific examples include:

### Over-decking

Over-decking by the provision of a secondary over-deck to relieve load on a deficient structure. One of the methods of over-decking is by commercial steel-polyurethane sandwich panels (Mylius, 2005). The polymer is sandwiched between the original deck and a new steel top plate. Wheel loads are distributed across the original deck and local stresses in the original deck are lowered. This method is particularly suitable for lightweight orthotropic steel decks which are susceptible to fatigue cracking of the welded connections beneath the deck plate.

### Deck thickening

Thickening the concrete deck of a composite bridge to enhance the load-carrying capacity. This can be achieved by the imposition of added concrete, suitably bonded to the existing deck, or it may be necessary to replace the existing deck altogether.

### Substitution of lightweight deck

Reduction in deck weight by substituting lightweight concrete in place of existing conventional concrete. This method is applicable to old bridges which have been repaired or strengthened in the past but continue to require remedial action. In some cases the concrete may have deteriorated due to corrosion of the reinforcement and require replacement irrespective of its load carrying capacity. A successful example of substitution of lightweight concrete is given in Appendix A2.1 on Coalport Bridge.

# 10　Remedial works – strengthening and replacement

> Remedial works and the various factors related to strengthening of structures are described. These include:
> - factors for selection and execution of strengthening techniques
> - specific methods and examples of strengthening
> - guidance on use of fibre reinforced polymer strengthening techniques
> - considerations relevant to the construction operations.

## 10.1　INTRODUCTION

Careful consideration of the strengthening or replacement methodology is important at all stages of the design process, particularly during the initial conceptual phase. Practical guidance on construction issues is given at the end of this chapter.

Strategies for strengthening include:

1　Reduction of the perceived load effects on affected components.

2　Increase the resistance of affected elements.

The following sections describe applications of each in more detail. Some examples of strengthening iron and steel bridges and other relevant sources of information are given in the case studies in Appendix A2.

In some cases, heritage considerations will dictate that the structure is preserved with as much as possible of the original material undisturbed when it might otherwise be more economical for it to be replaced. In other cases, the fabric of the existing structure may be retained in a purely cosmetic function, and the load carrying capacity provided by a modern structure hidden within the overall bridge envelope (see Tilly, 2002). The general requirements of heritage legislation are summarised in Sections 3.4 and 8.5.

## 10.2　STRENGTHENING BY REDUCING LOAD EFFECTS OR MODIFYING LOAD PATHS

In Section 8.4 a number of actions were described that can be used to review whether the loading effects are actually as calculated. These options should be explored fully before embarking upon remedial work to the structure and can include:

- observation of the structural behaviour (vertical deflection, horizontal movement at bearings and joints)
- strain gauging and deflection monitoring
- load testing to compare measured results with analysis
- measurement of temperature variation and thermal response of the structure.

As an alternative to providing additional strength at a critical section, the structure can be modified to change the way that load is distributed, and diverting load away from the critical areas. This strategy can be used with advantage where areas to be strengthened are difficult to access, or where there is insufficient room for the connection to ensure that additional material is properly mobilised.

While additional load paths can be used to improve the overall static strength of a structure, there are benefits in using this approach to improve the fatigue life of critical details. In the approach taken in BS 5400-10 (BSI, 1980), the fatigue life of most welded details is related to the third power of the stress range for lives up to $10^7$ cycles and the fifth power for lower stresses so that relatively small reductions in stress range can provide disproportionately large improvements in calculated fatigue life. It follows that the adjustments to the structure to provide these reductions in stress range can be quite small.

Diverting load effects away from a critical area can be achieved by providing additional plating, bracing, or tension/compression elements in such a way that the relative stiffness of the elements is altered. Under elastic behaviour, the loading follows the stiffest load path to the support. By providing a stiffer load path than the one on which the component is situated, some load effects can be diverted away from the component.

These techniques are illustrated by Box 10.1 of the Avonmouth Bridge and Box 10.2 using a description of how additional girders and cross-bracing were used on the docks crossing of Docklands Light Railway to divert loading away from the original twin girders. Further details of these schemes can be found in papers on the Avonmouth Bridge by Gill, *et al* (1994), Kennedy-Reid *et al* (2001) and Pilgrim and Pritchard (1990).

Note that to be effective, it is usually necessary and sometimes essential to provide some form of preloading of the new elements, so that the engineer can be confident that the existing and new structural elements are working together compositely. An extension of this process is to increase the level of preloading of the new elements so that elements of the existing structure are relieved.

**Box 10.1**     *Avonmouth Bridge – strengthening using alternative load paths and additional material*

Avonmouth Bridge is a multi-span 1400 m long twin steel box girder structure. On the approach spans, steel box girders of constant depth act compositely with a concrete deck. The main 174 m span over the River Avon and the two spans on either side have an orthotropic steel deck and steel box girders of varying depth. Following a traffic study and assessment it was decided to utilise the existing width of the structure to provide for an extra traffic lane in both directions. Strengthening measures also addressed remedial work identified during the assessment process.

The bridge was under construction at the time of the introduction of the *Interim design and workmanship rules* (the Merrison Report). As a result, a high standard of fabrication was achieved. For a number of critical elements (eg deck stiffeners), *in situ* measurements of imperfections were much lower than assumed in the codified design methods. Detailed site records of the material strengths of all components were available, and showed steel strengths significantly higher than codified assumptions. Through use of this knowledge, a considerable amount of strengthening work was avoided. Strengthening measures included:

- inclined Macalloy (tensioned) bars used within the main box girders to relieve the loading on the webs over the piers
- additional longitudinal stiffeners provided to the top and bottom flanges of the box girders
- additional plating provided to stiffen pier diaphragm plates and bearing stiffeners
- additional plating provided to thicken existing web plates
- concrete infill between existing bottom flange stiffeners placed adjacent to piers to create a composite (strengthened) flange
- the construction sequence was carefully managed to ensure that the additional tendons and plating were participating in load sharing after installation
- following extensive testing of half-scale models and 3D non-linear computer modelling, stiffening plates were added to the main support bearings at the piers.

After completion of the original railway in 1984, development of the Canary Wharf complex in 1985-1990 led to demands for increased capacity and frequency of trains. The viaducts are sensitive to live load effects and fatigue was a major consideration in the design.

Additional material was provided to reduce the stress range caused by an increase in the required size and loading of the rail vehicles.

On the docks crossing, two additional girders were provided outside the original twin girder composite viaduct to support the (static design) requirement for longer platforms. These additional girders were connected to the original girders by a composite concrete deck and full depth transverse cross-bracings. The deck and cross-bracings were such that between 10 and 15 per cent of the live loading from the light railway vehicles was transferred laterally to the new outer girders. Although this is a relatively modest proportion of the total live load stress, the calculated benefit to the fatigue life on the original girders was significant. The bracing arrangement made good use of the new outer girders for both static and fatigue considerations, and enabled the amount of material that had to be added to the original twin girders of the viaduct to be minimised. The strengthening schemes are shown in Figures 10.1 and 10.2.

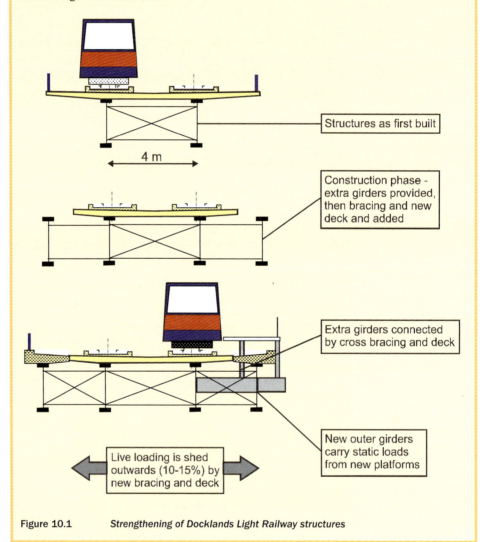

**Figure 10.1**     *Strengthening of Docklands Light Railway structures*

## 10.3     STRENGTHENING BY ADDING MATERIAL

### 10.3.1     Adding plates or sections

Typical elements requiring strengthening and suitable for added plating include flanges, webs, stiffeners and diaphragms. The strengthening may be required to address static strength or fatigue strength.

**Figure 10.2**     *Docklands Light Railway after strengthening*

## Improving static strength

If an element is stocky, and not prone to buckling, material may be added to enhance the static strength of the section in proportion to the ratio of required capacity to current capacity. Note however that addition of large amounts of material may necessitate some re-analysis of the structure, as the increased relative stiffness of a component will attract stress which may in turn lead to the need for further addition of material elsewhere. Examples are shown in Figures 10.3 to 10.6.

A further range of examples of strengthening by adding additional plating is contained in the publication by Kennedy-Reid *et al* (2001).

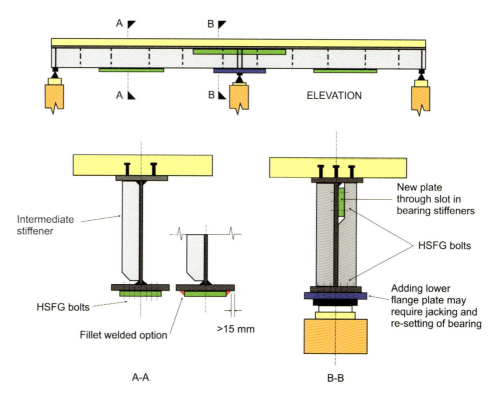

**Figure 10.3**     *Typical girder strengthening*

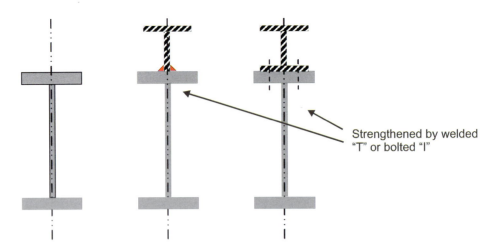

**Strengthened by welded
"T" or bolted "I"**

**Figure 10.4**    *"Top hat" strengthening*

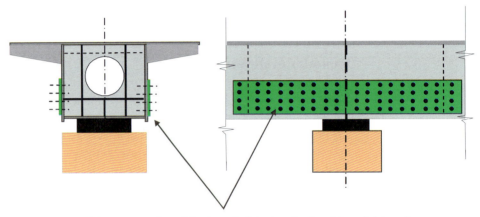

HSFG bolts, spaced to suit load transfer into strengthening plate, or (minimum)
as required by codified guidance for end or edge distance and spacing

**Figure 10.5**    *Pier strengthening for box girder*

## Improving fatigue strength

The calculated fatigue life of a component is usually expressed in terms of a number of
years, which can then be compared with the target life of the structure. In the UK, the
target life of a bridge is generally 120 years.

At a vulnerable detail, repetitive loading contributes to reducing the fatigue life of the
component and methods of identifying the stress ranges (cycles) are given in a number
of publications eg the reservoir method in BS 5400-10 (BSI, 1980). As explained in
Section 10.2 a reduction in the ranges of stress applied to a detail can produce a
significant and often disproportionate increase in its fatigue life. Additional material in
the form of plating will lower the stresses, as exemplified in Figure 10.6, and be cost-
effective. Note that the manner in which the additional material is provided and
connected to the structure can adversely affect the fatigue life if not done carefully. If in
doubt, specialist advice should be sought, see Section 10.3.4.

The fatigue life of a component can also be improved by reducing the frequency of the
loading or by diverting load away from the critical area using alternative load paths.

Local improvements of fatigue performances can be made by treatment of welds and
their profiles as described in Section 10.5.9.

The paper by Pilgrim and Pritchard (1990) describes how a mixture of additional material and alternative load paths were used to enhance the fatigue life of the original Docklands Light Railway viaducts to cater for increased frequency and magnitude of loading.

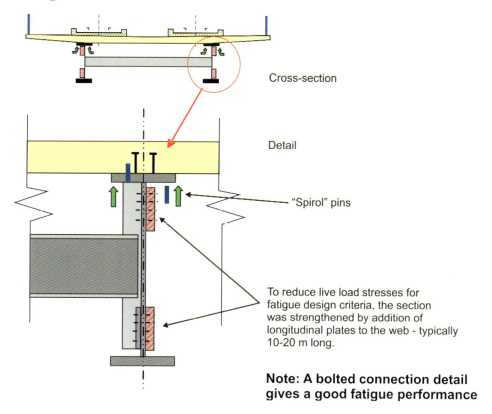

Cross-section

Detail

"Spirol" pins

To reduce live load stresses for fatigue design criteria, the section was strengthened by addition of longitudinal plates to the web - typically 10-20 m long.

**Note: A bolted connection detail gives a good fatigue performance**

**Figure 10.6**      *Strengthening by added plates and sections, Docklands Light Railway viaducts*

## 10.3.2    Stiffening

For structural elements that are in compression, and of slender proportions, buckling of the component can be a problem. The strength is limited by the buckling capacity rather than the yielding capacity. In these cases, it may be possible to strengthen the component by the addition of stiffening to the point where the buckling strength is higher than its yielding strength, making the best use possible of the material provided.

Examples of this include:

● subdividing large web panels on a plate or box girder with vertical stiffeners, to produce smaller panels

● subdividing large web panels with horizontal stiffeners, between vertical stiffeners

● providing stiffening to panels at points of concentrated load application (eg bearing stiffeners, local stiffeners)

● providing additional thickness of web or flange material

● providing discrete restraints to compression flanges of girders to restrict lateral or torsional movements, reducing effective lengths for buckling, and increasing static strength of the section.

Figure 10.7 shows methods of strengthening by addition of stiffening to improve the internal support zone of a continuous girder. The panel higher up the section will carry less compression and may actually be in global tension.

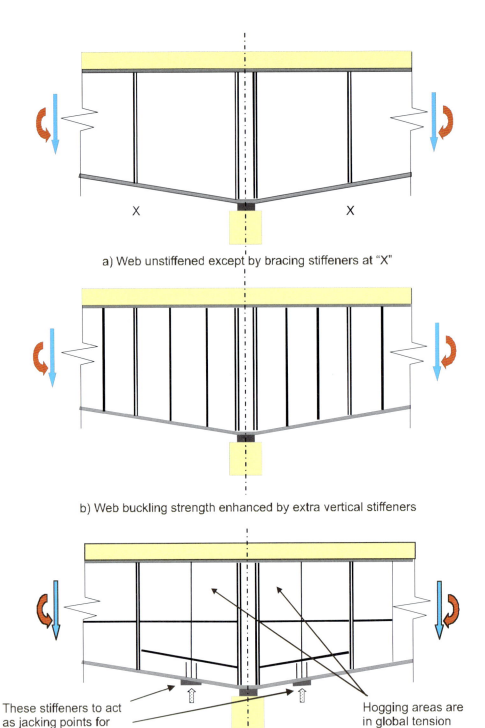

a) Web unstiffened except by bracing stiffeners at "X"

b) Web buckling strength enhanced by extra vertical stiffeners

These stiffeners to act as jacking points for bearing replacement

Hogging areas are in global tension

c) Web buckling strength enhanced by vertical and horizontal stiffeners

**Figure 10.7**    *Strengthening by web stiffening*

## 10.3.3    Connecting additional plating and stiffening

General principles regarding the addition of material are described in Section 8.4 and there is further discussion on forms of connector in Section 10.3.4. Bolting, welding, and increasingly specialist adhesives are used for connecting additional material to structures as part of repair or strengthening measures. Whatever connection method is used, it is important to consider the form of the detail. Ill considered and poorly specified details can result in repair or strengthening that is difficult to install or may result in the additional material not being fully mobilised, or worse, it may make the detail weaker than before strengthening was undertaken. Key considerations include:

1   Can existing unsound material be removed easily?

2   Is there room to prepare the structure properly (eg room for shot-blasting, drilling, welding preparation)?

3   Is there room for repair equipment (air wrenches for bolting, welding equipment, painting)?

4   Is the connection sufficiently long to ensure that the additional plating is fully mobilised at the critical section?

5   Are there enough connectors (bolts, welds) to transfer the required load effects?

6   Has the detail been arranged so that fatigue class (where relevant) is as good as, or better than, the original detail?

7   Will the repaired or strengthened detail require regular inspection after the work is complete?

Note that it is not enough to simply provide additional material at the critical section. Loads need to be fed into the additional plating some distance away from the critical section, in such a way that the new plating is fully mobilised at the location being strengthened. This is illustrated in Figure 10.8 where an example of a doubler (cover) plate is shown. Considering this detail from left to right, the load is fed into the doubler plate through welds or bolts connecting the plate to the underside of the flange, to the point where the two plates can be considered to be acting together. A similar transition needs to be provided beyond the critical section, to areas where the strengthening is not required.

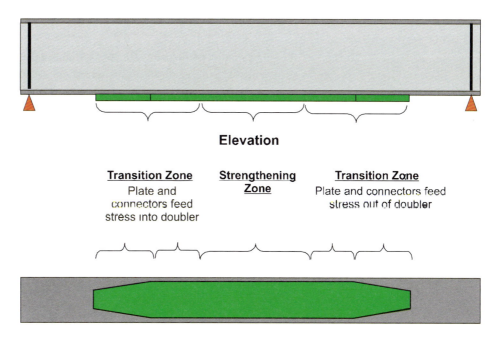

**Elevation**

**Transition Zone**
Plate and connectors feed stress into doubler

**Strengthening Zone**

**Transition Zone**
Plate and connectors feed stress out of doubler

**Underside of girder and doubler plate**

**Figure 10.8**       *Strengthening by doubler plates*

The length of the transitions can sometimes be considerable depending on the form of the structure, and the method of connecting the new plating to the existing structure.

### 10.3.4　Connections

Good detailing of the connection for installation and service is vital for a successful repair or strengthening operation. In this section, features of welding, bolting and shear studs are reviewed (rivets are considered in Section 9.3.7). Details on the use of adhesives are described in Section 10.4.

### Welding

Providing the parent material is suitable, welding is commonly used as it provides an excellent method of adding material to an existing structure. This should be established by testing or reference to material and structure data before the repair scheme is finalised. If in doubt, specialist advice on the suitability of the material should be sought.

Design should take into account that welding introduces significant amounts of localised heat into the structure. When the weld cools, the shrinkage will draw together the attached metal surfaces. Resistance to the shrinkage may result in large forces being developed. These forces can cause cracking in the weld, distortion of the detail around the weld, or distortion of the structure as a whole. If there are any discontinuities or defects in the material around the connection, these will be subject to forces that may make them worse.

If the material is suspected to have inclusions, or areas of lamination arising from the way in which it was originally made, applying a welded attachment to the surface may result in tearing or damage to the underlying material. Areas local to welded attachments should be checked beforehand for defects using ultrasonic or other forms of non-destructive testing.

The aim in designing welded connections should be to minimise the amount of heat input as far as possible by using an appropriate, but minimal, amount of welding.

There are other benefits that accrue from this approach as there is a limit to the size of weld that can be laid in one run. This depends upon the welding process and the welding position (downhand, vertically up, overhead etc). As an approximation, a 6 mm fillet weld can generally be laid downhand in one run. A 10 mm fillet will require extra care in relation to the amount of weld metal used, number of weld runs, heat input and resulting shrinkage effects.

An indication of the effects of shrinkage is given for the example of a fabricated T-section in Figure 10.9.

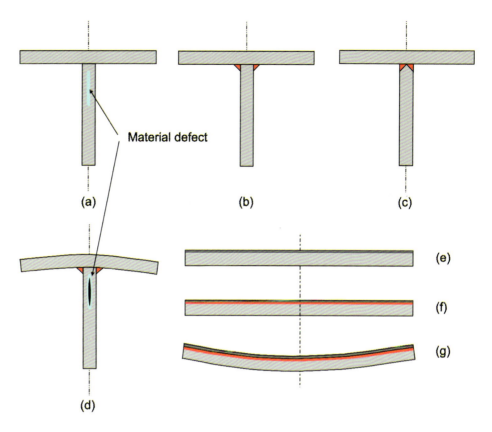

**Figure 10.9**     *Welding and weld shrinkage*

Diagram (a) shows the two 20 mm thick plates to be joined. An indicative (but much exaggerated) material defect is present in the material in the stem of the "T".

Diagram (b) shows a fillet welded connection. Depending on the level of loading in the T-connection, this could be as simple as a 6 mm fillet weld on either side of the stem From a fabrication point of view, this is preferred.

Diagram (c) is for the case when a very large fillet weld (say 10 mm leg length upwards) is required for strength and use of a butt weld should be considered. In this case, the additional expense of preparing the top edges of the stem plate by machining to a double "V" shape would be worthwhile as the resulting connection would use less weld metal than two large fillet welds. Heat input and shrinkage effects would be smaller.

Diagram (d) shows the effects of transverse shrinkage of the weld metal. As the weld metal cools, the flanges of the T-connection will be pulled inwards. If the restraint from the flanges is significant, the same shrinkage may well enlarge the defect within the metal. The weld metal also shrinks longitudinally.

Diagrams (e), (f) and (g) show a side view of the T-beam, with an indication of possible behaviour.

At (e) the plates have been prepared.

At (f) the weld has been completed.

At (g) the weld has shrunk, and has been resisted by the T-beam. The shrinkage has resulted in a compressive stress within the section and an overall curvature of the beam. If the plates of the T-beam were very slender, there could also be some danger of local buckling.

The technology of welding and weld shrinkage is complex. Guidance is given in SCI-P185 (SCI Steel Bridge Group, 2002). An appreciation of the principles involved is important. Each particular situation should be examined carefully, and advice sought from an experienced fabricator or welding engineer if necessary.

When considering welding, the following additional issues should be addressed:

1   Will the structure local to the repair be able to withstand the inevitable heating and restraint that result from the welding process?

2   Are there slender members located close to the repair that may buckle or distort?

3   If the structure is subject to repeated loads will the detail as strengthened have a fatigue class better or worse than the original structure?

4   Can the preparation of the metal and the welding be carried out in a safe working environment?

5   Is there room for good preparation of the area adjacent to the weld by shot blasting or other techniques? Note that the cleaning techniques should have provision for containment and collection of abrasive material.

6   Does the geometry of the joint allow sufficient room for the weld to be sufficiently carried out using the selected welding process?

7   Are the plates to be joined thick enough to require edge preparation and does this have to be done *in situ*?

8   Will preheating of the joint be necessary to avoid cracking in the weld area?

9   What trials are required of the procedure to be used and the *in situ* joints in production?

10   How will the welding process and consumables be protected from the weather?

11   What positions will be used for the welding?

   (a)   Downhand (preferable)

   (b)   Vertical (acceptable, but slower)

   (c)   Overhead (to be avoided if at all possible).

12   How will the joint be cleaned and protective treatment restored afterwards?

## Bolting

It should be recognised that a bolted connection is likely to require rather more physical space than a welded connection. The preparations for a bolted joint may be easier than for welding. Guidance on bolting is given in the *Steel designer's manual* (Owen and Knowles, 2003).

Bolts for structural use are generally of two generic types:

1   So called black bolts, which are of strength grade* 4.6 or 8.8.

2   High strength friction grip (HSFG) bolts, which are typically of grade 8.8 or higher.

The type of bolt and strength can be identified from markings on the head of the bolt and the nut, and each should be used with an appropriate grade of washer.

---

**Note**   * The strength grade of bolts is related to the values of UTS and yield stress. The first number is derived from UTS/100, and the second number from the ratio yield/UTS. So a 4.6 grade bolt has a UTS of 400 N/mm² and a yield stress of 240 N/mm².

The important difference between black bolts and HSFG bolts is that whereas black bolts will be installed and tightened as far as possible by the erector with a wrench, HSFG bolts require a predetermined torque to be applied to the shank of the bolt. This load is known as the proof load, and usually results in the minimum section of the bolt (often near the root of the threaded portion) being at or near the yield stress. Also, the plates are drawn so tightly together that local yielding occurs at high-spots.

Preloading of the HSFG bolts exerts a clamping action on the joint, and acts in a similar manner to the shrinkage forces in a riveted joint (see Section 9.3.7). The clamping action significantly improves the behaviour of the connection under fluctuating loads of vibration. The clamping action is beneficial in another way. Structural bolts are normally used in clearance holes (typically the diameter of the hole is 2–3 mm larger than the bolt shank), so there is the possibility of slippage between the plates of the connection. Depending on the geometry of the joint, such slippage may be a problem. The clamping action of HSFG bolts prevents slippage if properly designed and fitted. For strengthening work, the use of slip-resistant connections and preloaded bolts should always be considered, to avoid connections loosening under vibration or impact loads. The use of HSFG connections is recommended in preference to black bolts in bridgework (BCSA, 2005). BS EN 1993-1-8 (BSI, 2005) uses the terminology slip resistant connections and preloaded bolts.

There are two basic methods of installing HSFG bolts, both designed to ensure that the correct preloading is introduced to the bolt shank. These methods are known as:

1   Part-turn method.
2   Torque control method.

In the part-turn method, the nut of each bolt is tightened onto a hardened HSFG washer until a snug condition is achieved. On a large connection, several passes over the bolts may be needed to allow for bedding down. The nuts are then marked and an additional half or three-quarters turn is applied, usually using a pneumatic wrench or a torque multiplier and wrench. The thread of the nut is such that this is enough to yield the bolt shank and to impart the preload.

In the torque control method, a calibrated torque wrench is used to turn the nuts (from snug) until a predetermined torque is reached, at which stage the preload has been applied to the bolt.

Installation of HSFG bolts involves yielding, so bolts that have been installed once should not be re-used. Used bolts can sometimes be identified by the threads showing signs of stretching (the nut jams, and will not run freely). Both installation techniques require care, and guidance is given in SCI-P185 (SCI Steel Bridge Group, 2002). An example of an HSFG bolted connection is shown in Figure 10.10.

HSFG bolts are sometimes used with load-indicating washers. These washers have raised dimples around their circumference, which are flattened by the underside of the nut as the bolt is tightened. At the correct preload, a defined gap should exist between the surface of the washer and the underside of the nut. This can be measured with a feeler gauge.

There are a number of proprietary fasteners that are designed to operate in a similar manner to HSFG bolts. These include the Huck fastener, where the preload is applied by a jack acting directly into an extended shank of the fastener, and the nut is replaced with a collar that is mechanically swaged onto the shank when the correct preload has been applied.

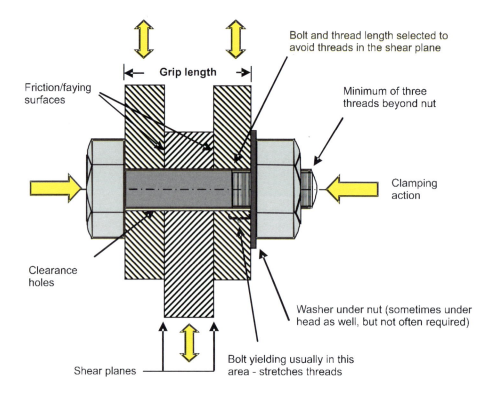

**Figure 10.10**  *Typical HSFG bolt installation*

The tension control bolt is another variation which requires a form of torque control using a special wrench that reacts against a stub extension of the bolt shank.

When considering bolting as a method of connection, the following issues should be addressed:

1    Is the connection designed to be slip resistant (recommended for all bridgework, unless for a temporary connection)?

2    Is there sufficient length of connection on either side of the critical section to provide enough bolts at the correct minimum spacing to develop the loads in the new (added) material?

3    Will the bolts use existing holes in the parent structure?

4    If not, is there room to drill the holes? Has the parent structure sufficient strength to withstand the loss of section from the additional holes during the installation process (perhaps with some control of loading)?

5    Is it necessary to consider a phased installation of bolts such that the number of holes without bolts installed is minimised?

6    Can the faying (contact) surfaces of the connected plates be accessed to achieve the correct surface preparation assumed in design?

7    Is there sufficient room around each bolt to allow use of an installation tool (air wrench or similar)?

8    Is the site such that quiet installation is required? A number of proprietary fasteners use electrically driven wrenches rather than air impact wrenches, and this has obvious benefits during night time operations in built-up areas.

## Shear studs

Welded shear studs are a feature of more recent bridge construction, and provide a relatively strong connection between steel member and concrete, commonly between the upper flanges of steel beams and a concrete deck to achieve composite action in new construction. Provided that the metal is weldable, and the metal surface is accessible, shear studs can provide a useful and flexible means of connecting to reinforced concrete to strengthen a member through increased composite action, or using the concrete to prevent buckling of a slender member.

To ensure good quality of the stud weld, the surface of the parent metal should be clean, dry and free from any contaminants that will affect the welding process, for example mill scale, paint products, galvanizing etc. In new construction, this is usually achieved as part of the normal fabrication process by shot-blasting of the girder. *In situ* stud welding, localised cleaning of the metal by abrasion may be required.

A range of shear studs is available. Some studs are threaded and can be used as the basis for a bolted attachment. Installation of shear studs, using handheld or automated equipment, can be swift. Key features of the process and terminology are shown in Figure 10.11.

In strengthening a composite structure having a concrete deck supported by steel girders, it is not often possible to access the upper flange of the girder to add shear studs without removing or coring the concrete. An alternative method of increasing the capacity by enhancing the strength of the shear connection can be achieved by using hardened steel Spirol pins, which were originally designed for use in mechanical engineering applications. This technique is shown in Figure 10.6, as applied to Docklands Light Railway viaducts in London, where an increase in both static and fatigue resistance of the structures was required to allow more frequent running of heavier trains.

Comparison of a shear stud with two typical types of pin is shown in Figure 10.12. The testing of the pins and application to the Docklands Light Railway viaducts is described by Pritchard (1992).

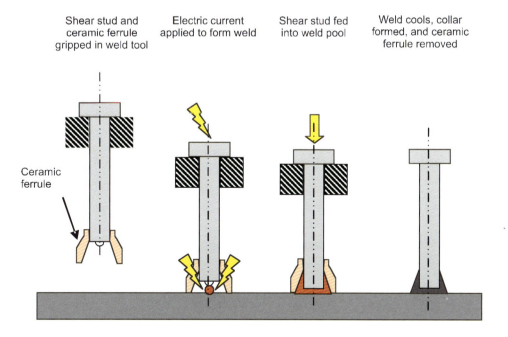

**Figure 10.11**       *Key features of shear stud welding*

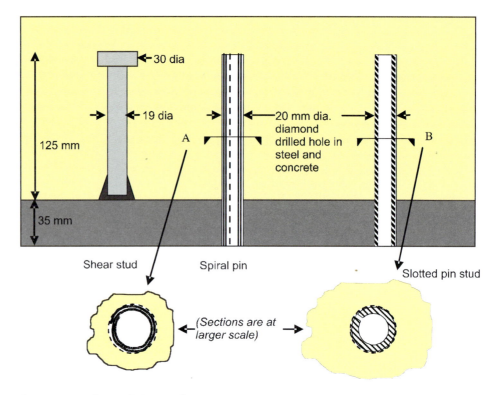

**Figure 10.12**    *Shear stud, spiral and slotted pin connectors*

# 10.4    FIBRE-REINFORCED POLYMER COMPOSITE STRENGTHENING

## 10.4.1    Introduction

In this section, a brief outline is given on the type of systems used for fibre reinforced polymer (FRP) composite strengthening and highlights some of the pitfalls that exist. A more detailed and comprehensive description of these techniques is given in *Strengthening metallic structures using externally bonded fibre-reinforced polymers* CIRIA publication C595 (Cadei *et al*, 2004).

FRP composites are very useful materials for repairing or strengthening iron and steel bridges. Typically they are bonded to the surface of the structure to enhance its strength or stiffness. Generally there are three main elements that comprise an FRP composite strengthening system:

1    High strength fibres such as carbon, aramid or glass.

2    Polymer matrix that binds the fibres together.

3    An adhesive that bonds the composite material to the structure.

FRP composites have good corrosion and fatigue resistance and have a high strength to weight ratio. Their use is particularly attractive where there are severe access constraints or high costs associated with installation time. They can offer significant advantages over conventional strengthening techniques as they can be rapidly installed and often the requirements for temporary works are significantly reduced. These advantages are a consequence of their high strength to weight ratio and often lead to significant reductions in the cost of strengthening. Furthermore they offer a non-invasive method of strengthening which negates the need for bolting or welding, which is a particular problem for cast iron bridges.

The main techniques used for FRP strengthening are as follows:

- prefabricated plates, manufactured either by a continuous moulding process (pultrusion) or preformed pre-impregnated (prepreg) sheets

- wet lay-up systems, made from prepreg sheets or woven fabrics

- vacuum infusion

- filament winding such as automated wrapping of columns.

## 10.4.2 Reasons for strengthening with FRP composites

A bridge might require strengthening by FRP composites to meet one or more structural deficiencies:

- lack of axial tension capacity

- lack of flexural tension capacity

- lack of shear capacity

- insufficient stiffness causing excessive deflection, inadequate buckling capacity or excessive dynamic response

- reduced fatigue life

- failure of connections.

FRP composites are particularly effective where their high tensile capacity can be utilised or to enhance the stiffness or buckling capacity of thin-walled structures.

It should be noted that at the time of writing, there is limited research material available for use of FRP composites to improve shear capacity.

## 10.4.3 Material types

### Cast iron

As cast iron has a low tensile strength it is often necessary to enhance the tensile capacity of the structural members. Due to the brittle nature of cast iron there is little or no redistribution of stresses at failure, so the strengthening is normally required to increase the elastic stiffness of the section, thereby reducing the elastic stress in the cast iron. As cast iron sections are generally larger than steel sections it is likely that significant areas of composite are required to reduce the stresses in the cast iron. For this reason ultra high modulus (UHM) carbon fibre plates are often specified. Alternative approaches to reduce the quantities of composite materials may include prestressing the FRP plates or stress-relief jacking whereby the bridge is jacked up prior to installation of the plates to reduce the permanent stresses in the structure. However, these approaches can complicate the anchorage stresses at the ends of the plates, and creep of the adhesive needs to be checked to avoid creep rupture of the resin or decay of the prestress. The magnitude of creep will depend on temperature and moisture as well as the permanent stresses.

### Wrought iron

Wrought iron is a ductile material so that a similar design approach to steel can be used. However, the possibility of delamination failure in the wrought iron needs to be considered and care should be taken when considering the use of CFRP to strengthen wrought iron members due to the risk of delamination failure and pull-off.

Wrought iron bridges are invariably of riveted construction which can cause added complications for CFRP composite strengthening as the rivets would interfere with the bonding.

### Steel

Steel is a ductile material exhibiting significant plastic flow at yield so that significant redistribution of the stresses can occur. Lower-modulus but higher strength CFRP composites are commonly used to strengthen steel bridges.

## 10.4.4      Types of FRP strengthening systems

### Prefabricated plates

Prefabricated plates are manufactured either by the pultrusion process or as preformed prepreg plates (prepreg is short for pre-impregnated fibres). Pultrusion is an automated manufacturing process that produces structural shapes of a constant section. The fibres are passed through a resin tank, before they are pulled through a heated dye and emerge as the cured and finished product having the fibres aligned along the length of the plates. The prepreg mats are laid up by hand to form a bespoke plate which is then cured under a high pressure and temperature. These plates tend to be more costly, but, unlike the pultruded plates, they can be manufactured with ultra high modulus carbon fibre, and to a precise design requirement. For instance, they can be tapered to significantly reduce the shear stresses in the adhesive at the ends of the plates and can be significantly thicker than the pultruded plates. Typically these plates are bonded to the bridge structure with a two-part epoxy adhesive system that cures at ambient temperatures.

### Wet lay-up systems

The wet lay-up process involves manually applying mats of FRP to the structure and then impregnating the FRP with a liquid resin system. A wet lay-up system may be consolidated by a vacuum bag. In this case a release film is applied over the composite, followed by a polymer membrane which is sealed along its edges. Air is then extracted from the bag so that a pressure is applied to the FRP. These systems can be applied to curved surfaces.

### Resin infusion techniques

Resin infusion is similar to a vacuum bag consolidated wet lay-up system, but in this instance dry fibre mats are applied to the structure. The fibre mats are then covered with a diffusion membrane followed by a vacuum bag. A vacuum is then used to draw in the resin and impregnate the fibre mats.

### *In situ* prepreg lamination

Prepreg fibre mats are applied to the structure, and covered in a vacuum bag and electric heating blankets. The FRP composite is then cured at an elevated temperature and under pressure. This approach along with the resin infusion technique, produce composites with a high volume of fibres resulting in high strength and stiffness properties. These processes tend to produce a higher quality FRP composite than the hand lamination techniques.

### Filament winding

The filament winding technique involves the automated winding of prepreg fibres around a component such as a bridge column. The prepreg fibres are typically cured at an elevated temperature to produce a high quality FRP composite. Filament winding techniques have been used to enhance the ductility and strength of concrete columns by confinement of the concrete and, less commonly, to strengthen metallic elements.

Filament winding is only likely to be cost effective if applied to many similar sized components.

## 10.4.5    FRP design issues

This section highlights some of the design issues for FRP composites that an engineer should be familiar with. It is not intended to give comprehensive coverage of the design, which is covered in more detail elsewhere in other publications. It should be noted that FRP composites have anisotropic material properties with significantly greater strength in the direction of the fibres. The design should confirm the strength and serviceability of the substrate material (the metal), the adhesive joint and the FRP composite. In most schemes the critical element will be the adhesive joint.

### Adhesive joint analysis

The concept of externally bonding FRP composites to metallic structures requires the transfer of high shear stresses across the adhesive joint to ensure composite action between the FRP and the substrate. It is important that a good and reliable bond is achieved. Very high local shear and peel stresses can develop in the adhesive joint at discontinuities. These discontinuities may occur at the end or edge of the strengthening material, at bond defects or where discontinuities such as cracks or joints occur in the substrate material.

An elastic stress analysis can be undertaken to evaluate the critical stresses at discontinuities. A significant amount of work has been undertaken to characterise these stresses for typical strengthening geometries and this is covered by Cadei *et al* (2004). However, it is worth noting that due to the plasticity of the adhesive, the high peak stresses in the adhesive will be relieved at these discontinuities, so an elastic analysis will overestimate the stresses. This shortcoming of the elastic approach can be overcome by comparing the stresses developed in test samples using a similar analysis, ensuring consistency between the limit state load effects and the design resistance of materials.

A more rigorous approach to the joint analysis may be undertaken using a fracture mechanics approach. However further research is still required to analyse thermal effects and some areas of changing geometries.

### Thermal effects

The coefficients of thermal expansion of the FRP composite and the metallic substrate can be significantly different. For instance ultra high modulus (UHM) carbon fibre has a negative coefficient of expansion, so that it contracts when heated. However the typical resins used within the composite matrix have high coefficients of expansion so the overall coefficient for the composite material is typically positive. The difference in the coefficient of thermal expansion between the FRP and substrate should be minimised by careful selection of the constituent materials and the volume fraction of fibres.

The difference in the coefficients of thermal expansion can cause significant shear stresses in the adhesive joint. It is also important to check the stresses in the substrate material, as it has been found that the constraint forces due to this mismatch in the coefficient of thermal expansion can lead to overstress of cast iron elements.

Many of the FRP composite systems use ambient cured two-part epoxy systems. These provide a good and durable bond of the FRP composite to the structure. However these materials have a glass transition temperature (Tg) which is typically between 55°C to 70°C. As the temperature of a polymer increases towards its glass transition temperature the polymer softens and the shear strength is dramatically reduced. It is important that the in-service temperature is considerably below the glass transition temperature. If the adhesive is cured at elevated temperatures the Tg can be increased.

### Corrosion

Galvanic corrosion can occur when metals and conducting non-metals come into contact. A current may flow between them causing the metal to corrode. The risk of corrosion depends on where the materials are ranked on the electro-potential series and as carbon is widely separated from both steel and iron, the risk of galvanic corrosion if the materials are in direct contact is high, see Table 5.1. While it is possible that the adhesive will provide an adequate layer of insulation, it is recommended that a glass fibre (or other insulating fibre) layer is placed between the metallic substrate and any CFRP strengthening system, to prevent galvanic corrosion.

When strengthening weathered iron or steel, corroded areas should be thoroughly cleaned, preferably by grit-blasting or water-jetting and surface grinding. Where possible, bonding to corroded areas should be avoided as corrosion can sometimes re-activate despite best efforts to clean the surface.

## 10.4.6      Installation

The use of FRP composites in the construction industry is growing steadily. However it is still a relatively new method of strengthening that relies heavily on the quality of the workmanship. It is crucial to the success of the strengthening that the installation is carried out by an experienced contractor with suitably trained and supervised staff. Quality control testing should be specified by the designer.

For the strengthening of metal bridges the adhesive joint is the critical aspect of the design so the quality of the surface preparation and application of the adhesive is very important. A good quality and durable bond can only be achieved if all loose material is removed and the substrate is thoroughly clean. This should be followed by suitable abrasion. It can be difficult to obtain good joints between the FRP composites and high alloy metals that oxidise rapidly, such as aluminium and stainless steel. In these cases it is necessary to use acid-etching, which must be followed by neutralisation of the etching products. In all cases it is recommended that a primer is used to inhibit any corrosion after the metallic substrate has been prepared. The primer should also be compatible with the selected FRP composite system. An example of a plate bonding installation is shown in Figure 10.13.

It is also important to note that ambient cure two-part epoxy adhesives will not cure at temperatures below 5°C. It may be necessary to provide heating to ensure this minimum temperature is achieved. This may be a significant issue for the strengthening of bridges if it is required to install the FRP composites during a night-time possession.

**Figure 10.13**    *Adhesive placed ready for plate bonding*

## 10.4.7    Monitoring

The first known use of FRP composites to strengthen metallic structures was in the early 1980s when fatigue cracks were repaired in the aluminium superstructure of Type 21 frigates with CFRP and epoxy patches. It has also been widely used for the repair of pipelines in the offshore industry. However it is a relatively new technology for the strengthening of metallic bridges, so it is important that regular inspections are undertaken to confirm the continuing serviceability of FRP composites. The most common method of inspection is by impact-echo (hammer tapping) to detect any areas where de-bonding may have occurred. It is also important to look for rust staining which provides evidence of corrosion of the iron or steel at the interface with the FRP. Ideally, these inspections should be at six month intervals. Cadei *et al* (2004) give further guidance on what should be looked for. There is also a further need to develop non-destructive test techniques such as thermography.

## 10.5      SPECIFIC METHODS OF STRENGTHENING

### 10.5.1      Bearing stiffeners

The publication of BS 5400-3 in the 1980s introduced the requirement that bearing stiffeners should be fitted at the support positions in a steel beam or box bridge span to prevent the occurrence of web buckling. When bearing stiffeners are not provided then the bridge would fail its assessment.

The design of an end panel has always required special attention since the boundary conditions along the transverse edges are different from those of an interior panel. When the end bay of a plate girder buckles and develops a membrane stress field before failing, the end post must possess sufficient flexural and axial stiffness to support the loading to which it is subjected. The results of early laboratory tests often showed that when end posts were weak and deformed significantly under load the ultimate load carrying capacity of the girders was much lower than that of similar girders which had been fitted with stronger end posts. Once bridge design engineers became fully conversant with utilising the shear strength of webs it was important that the end posts were correctly designed. The geometrical design of bridge bearings causes the centre of bearing support to be inset from the girder end post, so it was logical for the first transverse stiffener to be placed close to the end post (at the bearing support position) in order to prevent buckling of the end panel.

Load bearing stiffeners, whether located at the webs in steel girders or on diaphragms in steel boxes, are designed to ensure that they are strong enough to resist the full moments occurring from bearing eccentricities. Eccentricities of the axial force in the stiffeners can arise from a number of causes. These can include thermal expansion/contraction, bearing misalignment, beam rotation under load, substructure movement and shrinkage/creep of composite decks. The axial force in the stiffener will be at a maximum immediately above the bearing, with the force dissipating over the length of the stiffener as load is shed into the web.

The bridge code advises that the ends of a bearing stiffener should be adequately connected to both flanges, but only advises that it be closely fitted to the flange subject to the concentrated load (normally the bottom flange). Edges which are required to be fitted should be clearly indicated on the drawings.

Following the assessment of many old railway bridges it has been found that often web stiffeners at bearing support positions were not provided. The girder ends were only fitted with a closing stiffener at the extreme end. Present assessment calculations frequently show a lack of capacity to carry the in-service loading because inadequate stiffening at the girder ends allows lateral torsional buckling to occur.

At the Midland Links Motorway Viaducts the steel composite reinforced concrete decks of 15–27 m spans are supported by universal beams for spans up to 21 m and those above are supported by plate girders. Bearing stiffeners were not originally provided, the transverse forces on the deck ends being carried by 150 mm thick RC shear walls supporting the deck ends.

A review of web buckling capacity ratios was carried out and a departure from standards was obtained which permitted the omission of bearing stiffeners from the universal beams. However, the same review also indicated that the plate girders were inadequate to resist the buckling mode, and so bearing outstand stiffeners were welded in place at all plate girder ends. The fitting of the stiffeners caused some difficulty and this required careful supervision. Where the web was significantly bowed and would not straighten the resulting gaps were filled with epoxy mortar. Weld testing was undertaken using magnetic particle inspection (MPI) and ultrasonics. Traffic loading restrictions were invoked during the welding of these stiffeners to avoid overstress, and the welding was carried out during normal working hours.

## 10.5.2 Post-tensioning

The bending strength of bridge deck girders is sometimes found to be inadequate. An effective solution to overcoming this problem can be obtained by the use of post-tensioning to provide additional tensile capacity. Prestressing bars such as Macalloy bars or prestressing cables can be used to obtain the optimum solutions.

For most systems the post-tensioning is carried out by bolting or welding brackets to the girders, through which the bars or cables are passed. These can then be stressed to the required value of tension. The nuts are then tightened to lock in the tensile stress. An advantage of these systems is that further tensioning of the bars or cables can be carried out if necessary.

**Box 10.3**     *Rakewood Viaduct – strengthening using post tensioning*

Post-tensioning was carried out on Rakewood Viaduct, a six-span continuous viaduct of composite construction carrying the dual three lane M62 motorway through the Pennines in West Yorkshire (see Figure 10.14). It was designed to BS153 and as a consequence of the live loading increasing the bridge was assessed to relevant parts of BS 5400, together with HA departmental standards current at the time which showed that the bottom flanges of the main girders over the piers were heavily overstressed.

Figure 10.14      *Rakewood Viaduct*

The option of adding bottom flange plating at the piers was rejected because of the obstruction by the bearing stiffeners above and the bearings below. The solution was by prestressing bars fitted between anchorages, which were bolted to the underside of the bottom flanges using HSFG bolts. Bearing stiffeners also had to be fitted above the anchorages to resist the induced local vertical forces. This solution was advantageous because there were no restrictions on headroom beneath the bridge girders.

At Avonmouth Bridge, Macalloy bars and prestressing cables were fitted within the steel boxes in a coat hanger arrangement over the pier diaphragms to reduce the hogging moments over the piers and the shear stresses in the webs, see also Section 10.2.

At Friarton Bridge, prestressing cables were installed beneath the top flange of the twin boxes over the supports to relieve the top flange overstresses caused by increased traffic loading and current assessment standard requirements.

### 10.5.3      Concrete encasement

Following an assessment when steel members are found to be overstressed and liable to buckle, it is conventional to strengthen by the addition of steel plates and sections. However, there can be valid reasons for strengthening the overstressed member by using concrete encasement. Provided the additional weight does not give rise to other problems in the structure, the use of concrete can be shown to be beneficial and economic. The solution is particularly suitable if placed directly over the support where the additional weight of concrete is much less of a disadvantage.

An example of using concrete in this manner is when the longitudinal stiffeners located in the bottom flange of a box beam are found to fail their assessment in flexural bending over the pier supports. In one case a 9 m length of longitudinal stiffeners along the bottom flange in the region of the piers was encased in concrete to reduce the effective length, and thereby pass the assessment. This work proved to be more economical than strengthening the stiffeners with steel plating and avoided welding in a confined space. Using concrete within a box beam works well because the bottom flange, side webs and transverse bottom flange stiffeners are all able to act as formwork to the concrete placement. If required, longitudinal and transverse reinforcement, and shear studs can be used to complement the addition of concrete.

Concrete can also be used effectively inside overstressed closed box sections. For example, the addition of concrete has been used to overcome the assessed buckling failures in the columns of multi-level motorway interchange bridges, the compression strut box members of bowstring and arch bridges, and the steel box towers of cable stayed bridge spans.

Huntworth Viaduct, a 17-span twin steel box girder bridge carrying the M5 motorway over several different transport routes, was inspected and assessed following the failure of one of its steel roller bearings. It was when the bottom flange longitudinal stiffeners over the piers were found to fail that a strengthening method evolved using concrete encasement.

During the construction of the Erskine Bridge over the River Clyde, west of Glasgow, design calculations showed an inadequacy in the already built towers to resist the buckling of the closed box under the action of maximum cable down loads and the worst lateral wind loading case. The tower construction comprised a simple stiffened plate box which tapered from the base to the upper cable saddle. Wet concrete was pumped through pipework into the tower box, with additional lengths of pipework being fitted as the concrete level rose up the tower. Once the operation was complete the only means of access to the tower top was by outside hoist or crane, instead of by the original stairs fitted inside the box between platforms.

### 10.5.4      Replacing trough deck infill with reinforced concrete slab

There are many bridges with decks comprising metal trough decking supporting compacted granular fill and topped with a mastic asphalt carriageway surfacing. Assessment calculations of these bridges often show weaknesses in bending of the troughing. A well proven and suitable method of strengthening this form of construction has been to retain the troughing as permanent formwork, remove the granular fill and replace with a reinforced concrete deck capable of carrying the design loading, with the troughing no longer contributing to the capacity.

It is often inadvisable to attempt to weld additional strengthening flange plates on to the troughing due to the poor condition of the existing steel surface and doubts over long-term durability.

Argyle Street Bridge, Hull, has three steel spans with brick arch approach viaducts, see Figures 10.15 and 10.16. It carries Argyle Street over four rail tracks to the west of Hull Paragon Station. It comprises a three-span wrought iron superstructure with five main longitudinal girders supporting transverse trough decking. The deck is supported on brick abutments at each end that are integral with the approach viaducts. Two rows of five cast iron columns provide the intermediate support.

Figure 10.15     *Argyle Street Bridge, Hull*

The bridge was first inspected and then assessed. The BD 21/01 (HA, 2001) assessment identified several elements of the superstructure to be under strength, including the main carriageway and edge girders, and the transverse troughing and its supports. In order to achieve a 17T rating to BD 21/01 one of the strengthening options was to design a new transversely spanning reinforced concrete deck slab which used the existing metal troughing as permanent formwork. Other items of strengthening included new plating to both repair and strengthen the main girders, and the installation of Trief kerbs and P4 pedestrian barriers.

Figure 10.16     *Argyle Street Bridge during remedial work showing original troughing and infill material*

## 10.5.5     Infilling bridge spans

Sometimes it is necessary to consider infilling a bridge span so as to convert it to an embankment rather than attempting to strengthen the structure itself. Some typical conditions in which this type of infill scheme can be used are:

* when a span has reached the end of its effective service life, typically due to corrosion

* when the bridge support structure requires repair or strengthening, the cost of which is out of all proportion with the possible future upkeep of the bridge

* in a viaduct where it is possible to infill some of the end spans starting from the abutment support

* when the original road, track, footpath or water course is no longer used or required, or has been diverted.

A good example of such an infilling scheme is the Battersea Yard Railway Bridge in London, described in Appendix A2.6.

## 10.5.6     Additional beams

### Main longitudinal girders

Cast iron beams cannot be easily strengthened and one option is to install new steel beams between the existing beams. A suitable method needs to be evolved to ensure that the load on the cast iron beams can be transferred directly to the new beams. In the case of several bridges where this method has been successfully employed, it became necessary to install universal column (UC) sections with a reduced depth rather than a universal beam (UB) section. This then provides sufficient room to jack them against the deck soffit as well as ensuring enough space below the beam at the point of support to provide a suitable seating. Transverse steel sections can then be fitted between the existing and new beams to transfer the load from the cast iron beams to the steel beams.

Brandy Wharf Bridge in Lincolnshire, was strengthened by the addition of arched longitudinal steel beams inserted between the original cast iron beams which were left in place but relieved of loading.

### Transverse beams and bracings

Where cross-girders are found to be under strength they can be replaced by stronger girders which are seated on the existing main girders provided that these are shown to have adequate strength. In half through girder bridges, when the main outer girders are assessed and found to have weak connections to the transverse beams and cannot provide sufficient U-frame stiffness or strength, then it is necessary to strengthen the U-frame connection. Precise details of the type of strengthening will depend largely on the exact connection at the crossbeam ends. For bolted moment connections, bolt capacities can be increased by adding further splice cover plates which convert the bolts to double shear capacity. The capacity of welded moment connections could be increased by the addition of extra welded cover plates. In all these situations it is necessary to avoid strengthening existing bolted connections by the use of welds, and vice versa, in accordance with the requirements of BS 5400.

For those composite bridge decks where the main girders are sufficiently deep, then additional transverse bracing provides another strengthening solution if deficiencies in main girder stability are found during bridge assessment. In these cases, it should be ensured that the adjacent transverse web stiffeners or their connections are checked, as they may need to be strengthened in order to transmit the forces from the bracings.

## 10.5.7 Substituting alternative materials

In some situations it may be advantageous to replace original members with similar ones made from more modern materials having improved properties, for example higher strength, greater toughness, and better resistance to impact.

This method has been successfully carried out on occasions when it has been found that the strength of the original iron or steel members was insufficient. Ductile SG (spheroidal graphite) cast iron was used for replacement stanchions at the ends of crossbeams to strengthen the Clifton Suspension Bridge. This substitution has also been used to increase the resistance to impact of parapets on the Adelaide Bridge, Leamington, and on Westminster Bridge, London. Substitution has the considerable advantage that the new castings can be made from moulds copied from the originals so that the historic appearance is unchanged.

Main beams with inadequate strength can be replaced by similar sized steel beams having higher load carrying capacity and impact resistance. This method was used to strengthen Queen's Avenue Bridge, Farnborough. Here, seven original wrought iron plate girders were in poor condition and were replaced by twelve rolled steel beams.

For Holloway Road Bridge, described in Appendix A2.11, it was necessary to strengthen and reconfigure the deck to enable it to carry electric trams in 1908. Ten of the main cast iron beams were substituted by thirteen steel beams at spacing that varied to suit the loading requirements of the trams. Further repair and strengthening was required in 2005.

## 10.5.8 Parapet upgrading

The existing stock of metal bridges on motorways and other trunk roads exhibits a large variety of parapet types. Many of these are not adequate to cope with the needs of present traffic as expressed in the current criteria for the design of new bridge parapets. The superstructures of these bridges may not have sufficient capacity to transmit the impact forces from parapets of modern containment standards. Also they may not have sufficient reserves of dead load capacity to allow additional strengthening members to be added to the structure, and so partial rebuilding of the structure may be required.

Guidance regarding minimum containment levels for parapets may be found in the *Design manual for roads and bridges* (TSO, 2007). Before work can start, confirmation from the relevant road authority should be sought.

Before undertaking the strengthening of existing bridge parapets, the structure supporting the parapets should be checked to ensure that there is adequate strength for the additional loadings imposed due to the upgrading of bridge parapets. The impact forces on parapets are quite severe and can cause significant moments and shears to act locally at the post fixing, more widely in the deck cantilever supporting the parapet, and for high containment parapets globally on the deck structure, its bearing and support. Further strengthening members may be required to improve the strength of the deck cantilever and carry the moment back to the deck longitudinal members.

Various methods can be used to strengthen or replace parapets but for the older bridges the choice is often limited by the requirement to keep the appearance unchanged. This applies particularly to bridges having cast or wrought iron parapets. Where the parapets are of cast iron, the original material can be substituted by ductile SG cast iron. At the same time connections between segments and to the deck should be reviewed and strengthened as necessary.

On Magdalene Bridge, Cambridge, the parapet has been strengthened by fitting prestressing strand longitudinally tensioned between the pillars at each end. The pillars have been given reinforced concrete cores and are clad in stone. The strand was positioned so that it could not easily be seen and the appearance of the parapet, including the pillars was unchanged.

On several bridges the original parapets have been left unchanged and standard barriers erected on the edge of the footway to give protection to pedestrians as well as errant vehicles. In situations where the condition is sensitive, the barriers have been designed to have an appearance compatible with the age of the bridge as on Battersea Bridge across the Thames and Argyle Street Bridge, Hull, shown in Figure 10.17.

**Figure 10.17**     *Trief kerbs and barrier on Argyle Street Bridge, Hull*

As an alternative, or in addition to added barriers, built-up kerbs can be used. These can be constructed in SG cast iron or concrete and dished to redirect errant vehicles as on Cleveland Bridge, Bath.

## 10.5.9     Weld strengthening

In recent years there has been much research work and investigation into the application of weld improvement methods to increase the fatigue life of existing steel bridges, see Maddox (1991), Gurney (1991), Narayanan (1991). A large proportion of metal bridges, carrying either road or rail traffic, are highly stressed by cyclic and changing loads, and will reach their service lives in the coming years. However, the service life of those fatigue loaded steel structures is significantly influenced by the fatigue strength of critical notch type details, especially welded joints.

In addition some bridge girders fail their assessments through inadequate fillet weld strength and there is often the need to increase these in size to restore adequate capacity. Great care needs to be taken with any form of weld strengthening because of the possible weakening of the heat affected zone (HAZ). In all such cases the integrity of the existing welds needs to be established first. Checking the integrity of welds at critical locations is of importance by dye penetrant testing, magnetic particle inspection (MPI), ultrasonics or radiography. Often it is necessary to remove any protective coating before testing. More recent techniques, which are finding increased favour, are alternating current field measurement (ACFM) and acoustic emission. ACFM is a non-contact technique which is operable through coatings and relies on cracks in the welds disturbing a magnetic field for their disclosure.

While the joint design has a major effect on design life and is the basis for calculating service performance, the weld quality also has a decisive effect. Any fatigue analysis assumes that the welds are of an acceptable quality and comply with the inspection acceptance standards. In practice, of course, it is not always possible to guarantee a perfect weld and cracks, lack of fusion, slag entrapment and other planar defects may be present, significantly reducing the fatigue life. The common shapes of fillet welds are shown in Figure 10.18. Mitre is the theoretical shape assumed for design purpose, concave is the shape typically produced by improvement methods, and convex of an as-welded fillet.

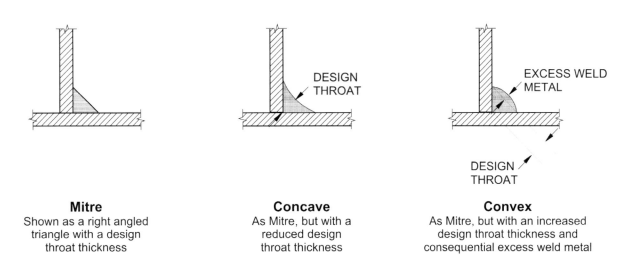

**Mitre**
Shown as a right angled triangle with a design throat thickness

**Concave**
As Mitre, but with a reduced design throat thickness

**Convex**
As Mitre, but with an increased design throat thickness and consequential excess weld metal

**Figure 10.18**     *Fillet weld shapes*

Weld improvement methods can be divided by their mode of functioning:

1    The first group produces geometric effects at the surface and includes grinding and TIG-dressing (Tungsten Inert Gas welding). These processes improve the weld geometry by smoothing the weld toe and reducing the notch radius. Burr grinding is applied on fatigue loaded structures such as bridges. It is common to use either hydraulic or electrical grinding machines to remove weld spatters as well as reducing microcracks and the sharpness of the weld toe. TIG-dressing also causes a reduction in the sharpness of the weld toe.

2    The second group, known as mechanical treatment methods, produces inherent compressive stresses and causes surface hardening. It includes shot peening, hammer peening and needle peening. Hammer and needle peening also deform the weld toe. Both effects can be combined through the use of ultrasonic impact treatment (UIT) and ultrasonic peening (UP).

Grinding, TIG-dressing, shot peening, hammer and needle peening are well investigated methods. Several published studies have shown that the fatigue strength of weld details can be increased by the application of these methods if cracks, starting from untreated areas, ie the root of fillet welds, can be prevented. Published data also prove that the S-N curves of weld improved details have reduced slopes, so that the improvement rates are even higher for lower stress levels.

**Table 10.1**      *Evaluation of weld improvement methods*

| Method | | Advantage | Disadvantage |
|---|---|---|---|
| **Grinding methods** | | | |
| General | | • relatively simple and easy to perform<br><br>• inexpensive<br><br>• gives large improvements | • mainly used for planar joints that can be expected to fail from the toe<br><br>• grinding techniques give a poor working environment (noise, dust)<br><br>• access to weld may be a limiting factor |
| Burr grinding | | • very slow<br><br>• expensive owing to high labour costs and high rate of tool wear | • large improvements to be expected for all types of welds |
| Disc grinding | | • very fast compared with burr grinding<br><br>• can cover large areas | • score marks give lower improvements than burr grinding<br><br>• improper use may introduce serious defects |
| **Remelting methods** | | | |
| General | | • large improvements are possible<br><br>• suitable for mechanisation | • operator needs special training |
| TIG dressing | | • small physical effort required<br><br>• inexpensive | • careful cleaning of weld and plate required<br><br>• high hardness may result in C-Mn steels due to low heat input |
| Plasma dressing | | • easy to perform owing to large weld pool | • lower hardness than TIG dressing<br><br>• heavy cumbersome equipment<br><br>• accessibility may limit use |
| **Residual stress methods** | | | |
| General | | • large improvements possible | • not suitable for low cycle fatigue applications |
| Hammer peening | | • very large improvements possible for poor quality welds<br><br>• simple inspection criterion (depth of groove >0.6 mm) | • limited to weld toe treatment only<br><br>• excessive peening may cause cracking |
| Shot peening | | • well developed procedures for small parts<br><br>• covers large areas<br><br>• simple methods for quality control<br><br>• improves resistance to stress corrosion cracking | • practical application to large scale structure not demonstrated<br><br>• best suited for mild notches<br><br>• very thin surface layer is deformed<br><br>• corrosion may quickly remove beneficial effects |

Hammer peening with a round nosed tool or needle gun peening gives very good results, although the noise produced may prevent their use. Shot peening can also be used to introduce compressive stresses at weld toes with equally good results. Specifically striking the metal imparts a stress at the point of impact, which results in strain-hardening of that area. Strain-hardening raises the elastic limit of a material into the plastic range without affecting its ultimate strength. A strain-hardened material will not deform under the same low stresses as a non-hardened material. When using peening as a method of weld improvement, care should be taken to ensure that any surface cracks present in the weld are detected beforehand. If the weld has not been checked previously using a suitable NDT method, peening will simply cover up the crack, and possibly drive the crack further.

While peening is ineffective for weld root cracks, hammer peening is the best method to repair surface cracks which are less than 3 mm deep. Also pre-peening can improve the detail by one fatigue class. Weld repairs can be expected to last about as long as the original welds. By using peening methods it is possible to help weld repairs exceed the life of the original weld.

Although grinding and TIG-dressing are not as beneficial as hammer peening of the weld toes they have the advantage of being more consistent and easier to control. These techniques rely upon dressing the weld toes to improve the shape and remove any intrusions.

The dressing may be carried out using a TIG or plasma-TIG torch which melts the region of the weld toe, providing a smooth blend between the weld face and the parent metal. Alternatively the toe may be dressed by the careful use of a disc grinder. However, for best results, the toe should be machined with a fine rotary burr. Care should be taken to ensure that the dressing removes no more than 0.5 mm depth of material, sufficient to give a smooth blend and remove any toe intrusion.

## 10.6 COMPLETE REPLACEMENT

### 10.6.1 General

For reasons of capacity or deterioration, complete replacement of a structure is sometimes necessary. The availability of large cranes and experience of lifting large and awkward loads in restricted possessions of road, rail or waterways enables complete replacement of small to medium size structures to be cost effective, particularly if traffic disruption is an issue.

When undertaking such operations, the condition of the existing bridge should be carefully ascertained, so that stability of the elements during the dismantling and removal process can be assessed. If components cannot be safely lifted in large assemblies, the structure may have to be cut up *in situ*, with consequential effects on the replacement programme.

The option for replacement should never be disregarded even for a structure which may initially appear to only need local repair. It is important that enough investigation and exposure of critical elements of the structure are undertaken prior to committing to a repair scheme.

A risk management process should be used to consider the possible increases in cost and time if additional deterioration is discovered during repair against the cost of more comprehensive investigation and inspection in the first place. The range of repair costs identified from the risk process can be compared with complete replacement, and an objective judgement made.

## 10.6.2 Considerations for craneage

### General

In this section the means of installing replacements and additional components are considered:

- use of large cranes
- use of modular wheeled units
- sliding/rotating/launching operations.

Further details are given in *Guide to the erection of steel bridges* (BCSA, 2005) and *Guide to the design of railway bridges* (Iles, 2004).

### Use of large cranes

If the use of a large crane is envisaged, it is important to contact crane operators in advance, preferably at the design stage. New machines are constantly being developed and enhanced capabilities may be available.

Crane operators will be aware of specialist techniques (eg tandem lifts) that can make the difference between a viable and a non-viable scheme. There is a trend towards contract lifts, where a crane operator plans and delivers the required lift. This can be attractive, and brings the skills of the crane operator early on in the planning stage.

### Constraints

There are various constraints that may give difficulty at a site where a large crane lift is intended:

- large utility installations (gas, electricity, water, telephone)
- operation on or near embankments
- lack of room for crane outrigger pads
- ground conditions that will not support outrigger loads
- long reach from crane site to installation position.

The reach and weight of the lift are important factors to be considered. Modern cranes have sophisticated overload protection that will result in the machine locking up if allowable combinations of weight and reach are exceeded. The manner in which the structure or element is to be lifted can have an impact on the design. The stability and strength of each component need to be checked at each stage of the lifting process. If necessary, additional temporary bracing may need to be added to the structure, or components strapped together in pairs to form a stable unit for lifting and installation.

The crane should be selected to suit the site. The footprint of the crane with outriggers extended is particularly important, as is room for rigging, de-rigging, and slewing the machine. Reinforced concrete pads are often needed to support the outrigger loads. Setting out of these pads is important as 0.5 m is a typical tolerance and dimensions outside this limit can make a difference.

From time to time, a situation arises where the crane planned for the lift is unavailable. A substitute crane may be offered. Note that even if the substitute is of theoretically larger capacity, it may not be suitable, as the footprint may not fit the outrigger pads. Checks should always be carried out.

Crane operation is vulnerable to the conditions on site on the day of the lift. Early advice should be sought, and allowance should be made for the risk of high wind.

Breakdown of the machine should be considered, and reassurance sought regarding back-up in terms of the availability of spares, fitters and support crews.

### Use of self-propelled modular vehicles

The availability of self-propelled modular vehicles has reshaped the way in which heavy components can be installed. The units are a good alternative to a heavy crane as they are unaffected by wind.

It is possible to assemble a large structure away from the installation site (up to 800 m) and to move the assembly into place using the vehicles. In this way, structural integrity can be achieved before installation, and the bridge deck installed in one piece.

When modular vehicles are being used it is important to check the intended route for buried structures. Although the loads are controlled axle-by-axle, the total load from a number of closely-spaced wheels can be significant.

### Use of sliding/rotating/launching for structure installation

Sliding, rotating and launching techniques are not as prevalent as in the past as the availability of large capacity cranes and modular trailers has reduced the need.

The processes are, however, still viable where bad ground conditions exist, where there is a constricted site, or where a bridge deck is being replaced on a one-for-one basis. These are specialist techniques that need consideration at the earliest possible stage in design. The installation is frequently (but not always) a bespoke solution for the site.

Structure components are moved on pre-prepared tracks using PTFE pads, ball bearings, rollers, or proprietary roller skates to reduced resistance to movement. Movement is usually achieved through use of hydraulic rams, hollow jacks which pull on prestressing strand, or occasionally, winches.

# 11    Protective systems

This chapter deals with the different protective systems used on past and present iron and steel bridges. It covers:

- historic development of paints including surface preparation and toxicity
- performance of coatings and expectation of maintenance life
- maintenance painting procedures
- modern coating systems
- treatment of structures; iron work, buried structures, galvanized and complex configurations
- weathering steel.

## 11.1    BACKGROUND

Challenges arise from maintenance and repair of protective coatings that are already in place on a structure. These coatings may be of considerable age, and may involve materials now considered to be dangerous to health and safety if removed.

New bridges fabricated from non-weathering grade steel will have metal coating, paint coating, or a combination of both applied to protect the material from corrosion. The metal coating may be applied by thermal metal spray, or by hot dip galvanized, depending on the coating thickness required. Fasteners such as bolts may be electroplated, sherardized, or hot dip galvanized. In some circumstances, paint coatings are applied over metal coatings to provide a so-called duplex protective system.

Requirements for new construction are typified by those outlined by Highways Agency and Network Rail in their respective specifications. Paint systems for new construction have evolved from those having five or six coats to using three or four coats, applied sequentially. A useful overview of corrosion protection for bridgeworks is provided in *Steelwork corrosion control* (Bayliss and Deacon, 2002)

Newer systems may not be compatible with the existing systems without specific formulation or surface preparation, and maintenance funds can be wasted through premature failure of incompatible systems. If in doubt, the services of an independent specialist should be sought, supported by feasibility trials where judged appropriate.

This chapter uses a number of terms that the reader may at first find unfamiliar. Outline definitions are given in Table 11.1, with further explanation in the relevant parts of the chapter.

**Table 11.1** *Protective coatings*

| Coating | Features |
|---|---|
| Primer coat (primer) | • applied directly to the cleaned metal surface<br>• wets the surface<br>• provides good adhesion for subsequent coats<br>• may contain anti-corrosion pigments<br>• thermal spray metal coatings will be sealed before priming |
| Stripe coats | • localised application of a coat to vulnerable details – corners, fixings, edges etc to ensure proper coverage<br>• see Figure 11.8 |
| Intermediate coats (undercoats) | • applied to build the total thickness and provide a barrier<br>• thicker the coat, longer the life (in general)<br>• enhance overall protection |
| Finish coats (finishes) | • provide required appearance<br>• provide surface resistance – first line of defence<br>• resist degradation by sunlight |

Paint coats need to be compatible with one another, and it is important to allow for appropriate timings between coats.

## 11.1.1 Early oil-based paints

It is important to have a clear understanding of the early oil-based paints as they are still present on many bridges and require continued maintenance involving over-coating or removal. Morcover, for historic bridges, heritage authorities may require the maintenance operations to use materials contemporary with the time of construction.

When the first iron bridges were fabricated and erected in the 18th century, oil-based paints were the only method of protecting the metal surfaces from corrosion. These were composed of natural materials with an oil binder (generally linseed oil) and pigments, which were powders mixed into the oil. This paste was thinned with a solvent, usually turpentine. Each painter produced a mix of hand prepared paint to meet requirements of the particular application, and took great pride in their work and the quality of finish.

These air drying oil-based materials had very good wetting properties for metallic surfaces and the ability to penetrate any porosity in cast and wrought iron. They were also tolerant of surface rusting, surface condensation, and various forms of soluble ferrous salt contaminants, both on the metallic surfaces and within the porosity of any pits.

The pigments in use at that time were usually lead-based, either red or white lead, and reacted with the original oil binders. These pigments, in the presence of moisture and atmospheric pollution, formed lead soaps within the matrix of the oil-based coatings, which were extremely effective in preventing the spread of corrosion. In particular, they prevented the damaging and destructive pitting corrosion, which would have been devastating to the safety of early structural sections. However, the toxic dangers of lead pigments were realised during the 1960s with the result that they were significantly reduced, and then phased out during the early 1970s.

These early oil-based coatings have proved extremely effective in protecting iron structures, as can now be seen in the remedial work on the famous High Level Rail Bridge across the River Tyne in Gateshead, which was opened in 1849. Some of the original oil-based coatings are still present and although subsequently protected by multi-coat overlaying paint, have performed well during a period in excess of 150 years.

Another example of the protective properties of the early oil-based paints was discovered during the centenary remedial painting of Tower Bridge in London, which was erected and painted during the 1880s, and opened in 1894. During the paintwork survey in 1993, the inside of the counterweights to the bascules were found to have two coats overall of red lead in linseed oil. However, on the tens of thousands of rivet heads, an extra brush-applied stripe coat, tinted with a carbon black pigment for colour contrast, was found to have provided extremely good corrosion protection (as good as a multi-coat system) when the refurbishment was undertaken on the bridge. It has lasted very well and is the first recorded use of stripe coats to provide edge protection.

Removal of some of the multi-coat systems, also at the time of the centenary refurbishment, revealed that, in certain easily accessible areas, up to 31 coats of paint had been applied over 99 years of maintenance work. A cross-section of a removed paint flake can be seen in Figure 11.1, where the 31 separate layers of paint were identified.

**Figure 11.1**

*Cross-section through a paint flake having 31 separate layers*

## 11.1.2        Ready-mixed paints

With the arrival of ready-mixed paints and the development of paint chemistry between the First and Second World Wars, paint chemists introduced formulations of resins and oils as well as driers, together with faster evaporating solvents. This phase of paint chemistry enabled oleoresinous coatings based on both linseed and tung oil, with synthetic resins based on phenol and the newer alkyds, to be incorporated into the paint formulations.

These advances in paint technology continued to use the concept of oil-based coatings but with faster drying and tougher, more water-resistant resins, which were then applied to new steel structures at the time, as well as to iron and steel structures coated with weathered oil-based paints. The old paints were over-coated with newer oleoresinous coatings. These sandwiched compatible coating layers have provided long-term protection to the iron and steel surfaces.

The range of oleoresinous paints, which combined the oils with a range of more weather resistant resins, also incorporated a wider selection of anti-corrosive pigments including chromates, phosphates and metallic pigments such as lead and aluminium to provide greater corrosion protection. An inert, natural pigment, micaceous iron oxide (MIO), was introduced into intermediate and finish coats. This plate-like pigment provided excellent barrier protection. Multi-coats could be applied relatively quickly both at the fabricator's yard and on-site. The later coatings, which superseded the oil-based paints, also had relatively good tolerance for surface rusting and ferrous salts.

Single-pack, quick-drying, chemically resistant coatings based on chlorinated and acrylated rubber were introduced during the 1960s but were phased out in the 1990s with the introduction of the Environmental Protection Act (1990).

With the further development of chemical synthetic coatings after the Second World War and with the arrival of a range of longer life chemical resins and pigments, including two and three pack mixed materials with a potential life of 25 to 40 years, the importance of preparing the steel surfaces became abundantly clear to the paint chemists.

A range of more surface tolerant two-pack epoxy coatings with re-coatable modified polyurethane topcoats was also developed. This system has become a common standard in both the Highways Agency 1900 series and the Network Rail RT98 specifications. However, there has been a tendency by technical sales representatives from paint companies to promote these coatings to all new and maintenance painting contracts as being the ultimate cure all. A number of major failures have shown that a detailed survey by a qualified paint technologist/paint surveyor (not a painting inspector however experienced) should be carried out to ensure that these coatings would be compatible, both in the short- and long-term, with the existing paint.

## 11.1.3     Methods of surface preparation

When steel bridges were first fabricated the production and timing of the fabrication programmes enabled the steel to lose its surface mill scale by natural weathering so that the coatings could be applied to a wire-brushed surface, which provided a sound base for both oil-based and oleoresinous paints.

With the introduction of abrasive blast-cleaning, with shot or grit-blasting at the fabricator's works or expendable grit-blasting on-site, it was initially thought that a visibly clean white metal surface would be adequate for the laboratory-developed, potentially long-life coatings. However, it became clear that there was a significant gap between the preparation and performance testing of coatings in accelerated laboratory tests and the actual preparation and performance of the coatings applied on-site or at the fabrication works, which were subjected to the rigors of natural weathering.

Following a number of premature failures of coatings before the target life of 25 year had been achieved, it was realised that invisible contaminants on the visibly white metal surface, in the form of soluble ferrous salts, needed to be removed prior to application of the long-life highly sophisticated complex coating materials. These soluble salts are tenacious and can only be removed by a combination of water and abrasive (wet abrasive blasting). The acceptable level of contamination should be specified and measured using a conductivity meter.

A series of British and ISO Standards have now been developed to establish a satisfactory clean condition of the steel surface with the appropriate surface profile after abrasive blasting.

The standard BS 7079 (BSI, 1989) an appendix to ISO 8501-1 (BSI, 2001, updated 2007) was introduced by members of the British representatives of the ISO committee because the photographs of standards in ISO 8501 (BSI, 2001, updated 2007) are based on sand-blasting of various grades of corroded steel and do not represent the standards achieved with expendable or other recycled abrasives such as chilled iron or steel grit. So it is important to specify the standard from the BS 7079 (BSI, 1989) supplement rather than the ISO standard, which reflects an old Swedish standard. Figures 11.2 and 11.3 illustrate use of mechanised wire brushing and needle gunning, which represent the mechanical preparation methods for the ISO standards.

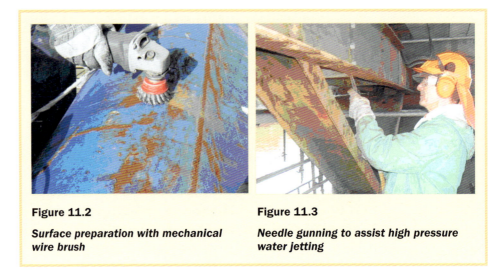

**Figure 11.2**

*Surface preparation with mechanical wire brush*

**Figure 11.3**

*Needle gunning to assist high pressure water jetting*

The new blast-cleaning methods and understanding the importance of removing surface contaminants, both visible and invisible, have resulted in a significant increase in the life of protective coatings on some of the newer steel bridges.

The introduction of blast-cleaning as a method of surface preparation enabled metal coatings, particularly thermally-sprayed zinc and aluminium, to be applied to the blast cleaned steel surfaces. These metal coatings were then over-coated with paint systems formulated to increase the protection of the metal coating.

### 11.1.4    Application skills

In more recent times, an increase in failures of coating systems has highlighted the drop in the standard of application skills. The original oil-based and oleoresinous coatings were applied by time-served painters, who had a real interest in the quality of their work. In addition they mixed their own oil, pigment and solvent to a consistency that suited their individual application methods and skills. However, during the second half of the twentieth century, with the arrival of ready mixed paints, less skillful labour was employed for application of the newer sophisticated coatings and there was less interest in their properties or the long-term performance of the applied film.

In order to improve the general quality of application skills, the Institute of Corrosion has set up a professional training and qualification scheme for painting operatives (ICATS), details are given in Appendix A3.

### 11.1.5    Metal coatings

The use of metal coatings, and the control of the formulation of paints to be applied both over metal coatings and direct to steel, were instigated by the former Bridges Engineering Standards (BES) division of the Department of Transport, now the Highways Agency (HA). A test programme was carried out which resulted in a series of paint systems being registered and incorporated into the Department's bridge painting specification. Initially, both zinc and aluminium metal spray were specified, but following some problems with thermally sprayed zinc, aluminium was chosen as the preferred metal to be used for thermally-sprayed metallic coatings. This blanket advice by the Department of Transport during the 1970s took no account of the success of the zinc system in practice, as noted below.

A protective system composed of thermally sprayed metallic zinc with a suitable sealer, followed by a paint formulation based on a tung oil/phenolic binder, pigmented with MIO, was applied to the Forth Road Bridge when it opened in 1964. This bridge received partial maintenance painting with chlorinated rubber based paints after approximately 20 years, but there are still areas where no chlorinated rubber has been applied and a major survey of the paintwork has established that the zinc metal spray and oleoresinous system has performed extremely well and lasted over 40 years. Recent detailed surveys of the paintwork, followed by feasibility trials, have shown that some of the zinc metal spray can be prepared and over-coated with two-pack systems to give a projected 40 year further life to the next major maintenance.

There have been some successes with hot-dip galvanized bridge beams, either over-coated with a duplex system or left in the zinc only state, for example, Stainsby Hall Bridge over the A174 adjacent to the A19 in Teesside. This bridge was examined after 30 years weathering and the zinc thicknesses measured during the survey showed that the loss of zinc was minimal.

Recently footbridges and pedestrian steel access ways have been hot-dip galvanized and then over-painted with a duplex coating system. There is, however, a general lack of understanding of painting onto hot-dip galvanized surfaces which has frequently resulted in paint failures due to basic malpractices, either in application or in the specification. The method of pre-treating the hot dip galvanizing, either when new, or after approximately 12 months' weathering, should be carefully specified

If it is decided to paint galvanized bridge structures, then careful selection of the coating system should be made and coating contractors, who have qualifications and experience of this type of application, should be selected to ensure that the optimum performance of the duplex system is achieved.

The example of Stainsby Hall Bridge has shown that galvanizing alone will last for 35 to 40 years and when a paint system is coated over the galvanizing, significant extensions to this life can be achieved. However, it is highly likely from an aesthetic point of view that painting would be requested before underlying base metal has been exposed, giving an initial projected life to first maintenance of 60 to 70 years.

## 11.1.6     Toxicity of coatings

The older oil-based paints and oleoresinous paints containing lead and other toxic pigmentation, including chromates, provide an environmental and safety problem when the time comes for them to be completely removed to enable a new modern coating system to be applied direct to the prepared base metal.

Toxic lead pigments were generally phased out towards the end of the 1970s and chromates partially phased out at the same time. However, there are a number of bridges where lead pigmented, oil-based coatings are still present. To minimise toxicity, these can be over-coated with compatible modern coatings containing milder solvent formulations.

## 11.2 PERFORMANCE OF COATINGS

The performance of protective coatings applied to steel structures since the 1970s, whether coated at works or on-site, has been extremely variable and there have been many dramatic failures as well as cases when the expected paint life has not been achieved. On the other hand, a major survey of over 1500 trunk and motorway steel bridges has shown that the coatings had a life to first maintenance painting of over 20 years (Deacon *et al*, 1998). This is in-line with the experience from a wide range of iron and steel bridges over the past 25 years, which has shown that when coatings are applied in accordance with a correctly detailed specification by experienced and competent operatives, they will last for at least 25 years between repainting cycles.

The reasons why bridge coatings fail to achieve their design performances fall into two categories:

1   The first category is the thermally-sprayed metallic-coated structures such as pedestrian footbridges, canal bridges and small railway bridges, where thermal spraying does not lend itself to give long-life protection to this type of design configuration.

    The reason behind this type of failure is that the numerous application passes, made by the metal spraying operator to build-up thickness, give extensive porosity within the metal coating layer. It is not possible to seal this porosity adequately, or indeed detect its presence, in difficult access areas. The metal spray on the steel surface acts like blotting paper, which sucks moisture and atmospheric contaminants through the applied paint film. This forms a corrosive poultice within the metal spray layer causing corrosion of the steel hidden behind the metal coating.

    An example of severe corrosion due to this problem is illustrated in Figure 11.4, where porous aluminium spray was inadequately sealed and four coats of chlorinated rubber paint did not protect the metal spray layer or the underlying steel.

**Figure 11.4**

*Example of severe corrosion of steel behind metal coating*

This hidden degradation is impossible to control without completely removing the failed paint and porous corroded metal spray. Otherwise, subsequent repainting would have to be at progressively shorter intervals with a life between each maintenance operation dropping from the target figure of 25 years to as low as 10 years.

Metal coatings on large flat or wide beam bridges and box girders, can be successful and achieve a life of 25 years provided the porosity in the sprayed metal coating is adequately sealed after its original application in the fabrication shop. Unfortunately, there are no available test methods for determining the adequacy of the sealing process.

2    The second category of failure is concerned with the paint-only systems where duplex coating systems have not been used. Often the failures are due to poor workmanship by operatives lacking the basic understanding of the sensitivities of the newer paints to environmental conditions and surface preparation standards. These types of failure are easy to repair, compared to metal sprayed coatings, but are costly and would be unnecessary if qualified and experienced operatives were employed and tight controls of application conditions were carried out during the maintenance painting.

Single-pack coatings including the range of non convertible coatings, such as vinyls, chlorinated rubbers and acrylated rubbers, that were used in the 1960s and 1970s before the arrival of more user friendly two-pack materials, have been phased out for environmental reasons.

Modern paint coatings have been formulated to perform satisfactorily for 25 to 40 years. This is confirmed by the success of the coatings applied to major bridges and structures like the gates of the Thames Barrier and the two steel-lined Mersey and Dartford Tunnels, where one-coat systems of hot applied solvent-free epoxy paint were applied and have now achieved lives in excess of 25 years. Figure 11.5 shows the typical condition of the coating of one of the main gates following water jetting to clean off dirt at the Thames Barrier, after 26 years.

Figure 11.5    *Thames Barrier main gate after 26 years immersion*

The Thames Barrier coatings have been surveyed after 26 years immersed in the River Thames and are now projected to last for at least 40 years before the first major maintenance is required. This demonstrates that the potential life of coatings can be achieved with the newer systems if all of the relevant factors are taken into account. The good performances of coatings on these structures were achieved by selection of the individual operatives who prepared the surfaces and applied the paint, and having full time qualified painting inspectors present during the critical stages of the workmanship.

## 11.3    MAINTENANCE PAINTING PROCEDURES

A series of steps should be taken before a decision is made on when and how a bridge structure should be repainted.

## 11.3.1      Paint survey of existing structures

A detailed inspection should be carried out by a qualified paint technologist, or experienced paint surveyor, to identify the condition of the existing paintwork in terms of adhesion to the base metal, cohesion, inter-coat adhesion and any unusual flaws in the coating. Flaws may be due to adulteration of the original formulation of the coating or adverse atmospheric conditions during the application or the drying/curing period. Any or all of these defects even in minor form, may adversely affect subsequent over-coating of the existing paint system when maintenance painting is required.

The Highways Agency documentation calls for a detailed survey of a structure for all Category III or IV failures and these surveys should be undertaken by a qualified paint technologist. A painting inspector qualified in supervising and inspecting the application of wet paint to properly prepared surfaces, is not necessarily experienced in evaluating the degradation of an existing coating, or understanding the changes that occur in the weathered paint systems, whether at the steel surface or within the body of multiple layers of paint coatings.

A paint survey should be undertaken within arm's length of the surface and a series of micro-destructive tests should be carried out:

- adhesion, by the St Andrews cross-cut test

- abrasion to bare steel

- cross-cut test of adhesion and cohesion

- thickness measurements of all layers by the paint inspection gauge (PIG test)

- solvent tests

- removal of paint flakes for microscopic examination (see Section 11.3.2).

Examples of micro-destructive test patches and sites of the other tests are illustrated in Figure 11.6.

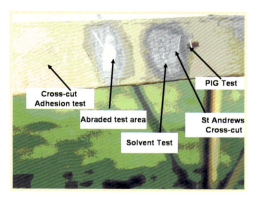

**Figure 11.6**

*Micro-destructive test patches*

This testing can establish the effects of a range of solvents in droplet form to assess the compatibility of the wide range of modern paint coatings, which could be considered for over-coating the existing paint system.

## 11.3.2      Laboratory testing of paint flakes

Small paint samples should be removed from various locations during the survey to be assessed under microscopy in the laboratory, as shown in cross-section in Figure 11.1.

The detailed laboratory examination of the paint flakes can also confirm the presence of toxic pigments and any evidence of unusual cohesive weakness or other premature degradation factors of the coating system.

### 11.3.3   Draft repainting specification

Following the detailed survey and assessment of the condition of the existing coatings, a comprehensive draft repainting specification should be prepared which is practical, technically sound, and having clauses specific to the structure in question. The draft can be finalised after it has been tested by feasibility trials. There is a range of methods and standards of surface preparation, which can be adopted for a bridge structure, to ensure that the right surface preparation standard (in terms of technical quality, long-life and commercial reality), is included in the document. On occasions when paint failures have occurred, it has invariably been found that the specification had been unclear to the operatives and, in some cases, the engineer responsible for the maintenance repainting contract. It is important to remember that a repainting specification should "say what it means and mean what it says".

The removal of toxic pigments in the early formulation of paint is a costly process. However, in a number of instances, the existing coating can be preserved by high pressure water jetting of the defective outer coats and surface contamination, and leaving the sound but toxic lead-based oleoresinous material intact, provided it is cohesively strong. A series of re-coating contracts was undertaken on rail bridges in central London during 2004 to 2006 using this controlled water-jetting technique (see Figure 11.7).

**Figure 11.7**
*Ultra high pressure water-jetting*

After water-jetting, a compatible single-pack paint system can be used to over-coat and preserve the life of the sound base material and the retained paint.

In any repainting specification, there should be specific clauses to cover areas of contact between non-metallic materials such as timber decks or concrete abutments where there can be difficulty with maintenance painting and accelerated corrosion can occur. Older structures having back-to-back angles, thin narrow plies and riveted sections should have specific clauses specifying methods of preparation and treatment followed by the use of a compatible sealant. If these guidelines are followed structures can be preserved and maintained for many years to come.

### 11.3.4   Stripe coating

The conventional application methods of coatings, whether by brush or spray, do not result in a full coating thickness on vulnerable points, such as edges, corners, welds, fixings. It is necessary to apply an extra coat by brush to these areas, as shown in Figure 11.8. This coat is known as a stripe coat.

It is essential to specify and ensure that a stripe coat is brush applied to these vulnerable points to build-up the minimum specified thicknesses because they cannot be measured with equipment such as a non-destructive electro magnetic film thickness gauge.

**Figure 11.8**    *Stripe coating of rivets*

Stripe coating as stated earlier, was successfully used in 1894 on the Tower Bridge rivets and should be used, with modern coatings on all edges, corners, plies, drainage holes and welds. This should be specified as an extra brush applied coat and not permitted to be applied wet on wet.

## 11.3.5    Feasibility trials

It is disappointing that many paint manufacturers who supply these modern coatings will advise engineers that their systems can over-coat an existing paint and upgrade it to a two-pack material. A number of major failures have occurred as a result of ill-advice. Engineers are recommended that a procedure (such as that of the Highways Agency) for surveying by a competent and qualified technologist, followed by feasibility trials to confirm that the draft repainting specification can be implemented, should be followed for all major structures.

A feasibility trial assists in confirming the practicality and adequacy of the draft specification as regards surface preparation, paint application and compatibility. This allows the draft specification to be finalised, as an accurate contract document.

## 11.4    MODERN COATING SYSTEMS

The development of paint coatings over the last 150 years, from oil-based to resin formulations, has left many old structures with a variety of coatings, which ideally should be suitable to be over-coated with modern paints. Unfortunately this is often not the case.

### 11.4.1      Surface tolerant epoxy primers

Surface tolerant epoxy primers have been developed since the late 1980s to have a degree of tolerance to surface moisture and adherent passive rust. However, numerous failures have resulted from misuse of these products where they have been applied to surfaces that have been too wet, too rusty or contaminated with salt. They may also be used over some aged but sound single-pack materials to upgrade them to a two-pack epoxy system but the feasibility should be established by a survey.

### 11.4.2      Glass flake pigmented coatings

Glass flake ideally should be a lamellar pigment having properties similar to MIO, with a carefully defined range and distribution of particle sizes, which when incorporated into a resin binder can provide flexibility to matured coatings and reduce stress during changes of temperature and humidity. Glass flake coatings also allow overlapping of the lamella flakes during curing, to provide greater impermeability to the ingress of moisture and atmospheric contaminants through the coating layer. Carefully formulated coatings containing the correct amount and size of glass flake pigment based on a two-pack epoxy, or a polyester resin binder, can provide a potentially long-life in excess of 25 years.

There are however, a number of so called glass flake or glass filled products on the market, where the glass pigment is little more than the addition of a small percentage of glass powder. This contributes nothing to the performance of the resinous material holding this pigment together.

Now that the importance of surface preparation is understood, the quality required for thermal metal spray should also be achieved for paint coatings. It follows that with modern paint systems it should be possible to achieve a 40-year life between maintenance coatings.

It is important that the outer layer of the selected coating be resistant to ultraviolet (UV) light, freeze/thaw cycles, rain, atmospheric condensation and contaminants, and should be in a condition to be re-coated at the time of the first major maintenance without the costly process of removing all of the degraded coating back to the bare steel.

If both the surface preparation and application of the coatings selected for the specific structure are carried out to a high standard by qualified operatives, then modern coating systems should resist the ravages of winter and summer weather, and any marine or industrial pollutants in the environment.

### 11.4.3      Moisture cured urethane

Moisture cured urethanes have been developed since the 1980s and comprise a range of chemically modified formulations, which complete the curing mechanism in the presence of atmospheric moisture. These products have a number of potential limitations in relation to their storage and performance which have restricted their use in practice.

### 11.4.4      Finish coats

A range of finish coats have been developed which will provide long-term protection over sound high-build and well formulated primers and mid-coats. These are commonly known as re-coatable urethane finishes.

The application of a relatively thin finish coat designed to be re-coated at the first major maintenance, is now becoming standard and is included in both the Highways Agency and the Network Rail paint systems.

Epoxy systems on their own will chalk readily when exposed to UV light and this combined with frequent atmospheric condensation and frost will result in an epoxy coating degrading to a point where it may need to be removed after 20 to 30 years or be in a condition unsuitable to be over-painted.

Polysiloxane coatings, which have greater potential life to first major maintenance, have been used successfully in the oil industry where structures with a design life of 20 to 30 years are coated without requiring any major maintenance during their lifetimes. It is not yet proven how over-coating of weathered polysiloxane systems would perform after about 25 to 40 years exposure but work on these products is continuing by a number of major coating suppliers.

Degraded coatings on old structures would not be suitable for over-coating with the polysiloxane formulations, unless all of the original paints are removed back to bare metal. This inevitably involves the removal of toxic lead pigments from the pre-1970s coatings.

Patch repainting of visibly corroded areas can be carried out with appropriate methods of surface preparation and the use of surface tolerant epoxy primers. Sometimes these coatings can be used as an over-coat for some aged oleoresinous paint systems. Where incompatibility occurs, single-pack coatings can be used for applications of this type, eg silicone alkyd or if available, vinyl type coatings.

If correctly coated at the time of construction, using best practice, repainting would only be needed three times in a lifetime of 120 years (the assessed life of highway bridges in BS 5400).

## 11.4.5      Thermal metal spray

There have been a number of developments of metal coatings, in particular the electric arc application of thermal metal spray (either zinc, aluminium or its alloys). These can now be applied as much denser and less porous coatings but the finished surface prior to application of the sealer and build-coats needs to be specified separately. To obtain long-life, the thermal metal spray should be carefully applied and only selected for large, flat beam surfaces with adequate sealing.

## 11.4.6      Hot dip galvanizing

The use of hot dip galvanizing has been mentioned earlier as a long-life protective coating for steel bridge beams. The main difference between this and other methods of coating steel is the bonding of the zinc to the steel surface. This alloying process results in extremely good adhesion which, coupled with the predictable rate of zinc loss, enables the time to first maintenance be predicted with some degree of accuracy.

The zinc thickness would normally be around 85 microns on steel sections that are at least 4 mm thick. This thickness can be increased to 125 to 130 microns with initial blast cleaning to increase the surface profile prior to galvanizing. The rate of loss of zinc in salt laden environments (de-icing salts during the winter months or a marine environment), is up to four or five microns per year. However, in rural or mildly industrial environments, the loss of zinc would typically be 1–1.5 microns per year and at this rate of zinc loss, 100 years could be achieved before repainting of the weathered zinc surface would be necessary.

As the galvanized surface degrades, there is a change of colour which is clearly discernable during visual inspections. This enables progressive thickness measurements to be recorded and predictions made as to when only the light brown zinc iron alloy layer will remain. This can provide an ideal surface for the application of the first paint coating, whether it is 30, 70 or even 100 years after the galvanized structure was first erected.

Coatings including direct application primers have been developed for painting either new galvanized or weathered zinc if the light grey, matt grey, dark grey colour of the slowly degrading zinc surface is unacceptable. In addition to the direct application primers, there is a new water-based environmentally friendly surface treatment (Actan), which can be over-coated with both one or two-pack systems to give an attractive appearance for many years.

## 11.5 TREATMENT OF SPECIFIC STRUCTURES

### 11.5.1 Iron structures

The maintenance and re-coating of cast or wrought iron structures should not be carried out in the same way as steel.

The treatment of iron structures will vary but the existing coating and any surface rust or scale should be removed before an examination is made of the condition of the ironwork. Removal of the original oil-based or oleoresinous coatings could reveal corrosion due to neglect or incorrect specifications being used over the years. Porosity can be found in the body of the wrought or more likely cast iron, but can be treated by grinding. In particular omega shaped pits or blow holes should be treated either by discing and/or grinding to ensure that the width of the neck of the opening is greater than the width of the base of the cavity. In this way, soluble salts or graphitic corrosion that has occurred over the years can be removed to enable sound paint performance and long-term protection.

Thin holding primers, such as two-pack epoxy prefabrication primers developed for steelwork, can be used to provide coatings which will penetrate the porosity of the iron surface. Once this penetration has been achieved, then thicker coatings composed of a primer, build coat and finish coat can be used to build-up barrier protection and give long-life to aged iron structures. It is important to have a tailored specification for repainting ironwork and not rely on a document prepared for steel.

The treatment of historic iron structures is often dictated by heritage issues as exemplified by the refurbishment of Pontcysyllte Aqueduct summarised in Appendix A2.7.

### 11.5.2 Buried structures

The use of the traditional bitumens and coal tar epoxy systems to protect buried and immersed sections of bridge structures has been ruled out on environmental health grounds since they have been classified as being carcinogenic.

Epoxies with exceptional flexibility, have been developed and the paint industry now refer to these as flexibilized epoxies. Early experience has shown that a number of these formulations can give excellent protection in areas where coal tar epoxy would have been used in the past.

### 11.5.3 Complex configurations

Bridges having complex configurations such as closely spaced cross-bracing and small section steel (for example lattice girders), could in theory be metal sprayed, but this is unlikely to be successful for an continuing life in excess of 40 years. This problem has been found on a number of structures including the Forth Road Bridge where zinc has been applied and on some motorway bridges where aluminium spray has been applied to thin cross-bracing sections. These cannot be sealed adequately and will inevitably contain high degrees of porous metal coating which will break down prematurely (see also Section 11.2).

### 11.5.4 Non-ferrous structures

Non-ferrous metallic materials such as aluminium alloy also need to be painted in certain corrosive environments as corrosion can, and does occur (see Appendix A1). As with iron and steel structures, the design and detailing of an aluminium alloy structure are important since ponding and areas with difficult access can lead to inadequate maintenance and accelerated corrosion.

Painting of aluminium bridge structures is relatively straightforward, but the specification for treatment, coating selection and application requirements, should be carefully drawn up and should not use a standard iron or steel specification. Two-pack phosphoric based etch primers provide the key to bonding the paint system to aluminium but this is moisture sensitive and needs to be applied under controlled conditions. Alternatively, a single-pack reinforced etch primer, with low acid content, could be used for surfaces that have weathered. Again, these etch primers can be over-coated with a wide range of modern coatings.

## 11.6 WEATHERING STEEL

Weathering steel has alloying constituents that enable it to develop a form of rust that has good adherence to the base metal, and in effect provides a protective coating. When first introduced it was claimed that weathering steel had resistance to corrosion per se and would not require painting during the foreseeable life of a structure. However, research by TRL based on long-term exposure to natural weathering at a range of test sites showed that this was not so, and it became apparent that it performed poorly in marine environments, certain industrial environments and continuously wet or damp conditions. In other more favourable environments there is a low level of corrosion and BD 7/01 (HA, 2001) recommends allowances of 0.5 to 1.5 mm on the thickness of steel members, depending on the environment.

In cases when weathering steel does not perform satisfactorily and remedial work is required, the first action should be to identify the cause of the problem. Poor performances may be due to:

- poor detailing enabling build-up of detritus permitting continuous wetting to occur

- local leakage of surface water containing contaminants such as de-icing salt

- change in the local environment since construction, for instance the construction of an industrial plant nearby.

It may be feasible to eliminate the cause of the problem, for example by improved water management. Otherwise it may be necessary to abandon the concept of weathering and resort to added protection by painting the steel.

## 11.7    ACTIONS TO ACHIEVE long-life

The initial protection and long-term maintenance of bridges with an assessed design life of 120 years should be taken as seriously as the overall design of the structure. Long-term performance has been improved by recent developments and an understanding of the requirements for surface preparation and coatings. It is evident that engineers should allow a 25 year maintenance recycling programme as a minimum objective.

The initial design of the structure should eliminate any features which would encourage accelerated corrosion or be difficult to maintain over the life of the bridge. This means that all surfaces on the structure should be accessible for surface preparation and application requirements in the future.

The design of the new structure should also be considered when the coating system is being selected, and this may exclude thermal metal spray on thin configurations or areas that may need multi-passes to obtain the specified thickness of metal coating. The areas which are not metal-sprayed, should have extra coats of the paint that are applied over other areas of metal spray. On completion of new construction or maintenance work, care should be taken to ensure that minor defects, such as construction damage or contact points have been repaired at least to the same standard as the remainder of the structure.

Maintenance of existing structures having unknown types of paint which have weathered, is far more difficult than for the more straightforward new construction. The first part of a successful maintenance painting contract is the detailed survey to identify the condition of the existing coating and to decide whether it can provide a satisfactory base for newer types of paint systems, which are likely to be more highly stressed than the original coatings or whether the existing coating should be removed back to bare metal.

Once the decision on the scope of the maintenance repainting contract has been decided, the coating system can be selected and a tailored draft specification should be prepared. This should be followed by feasibility trials before the final specification is issued. With both Highways Agency and Network Rail structures, the relevant systems and clauses can be adopted but specific features should be highlighted for every structure.

It is clearly important, for successful maintenance contracts, that the contractor and the painters employed on the contract have relevant professional qualifications, see Appendix A3.

If these basic principles are followed, the protection of iron and steel bridges will be successful in meeting long-life performance requirements.

# 12 REFERENCES

ADDIS, B and TALBOT, R (2001)
*Sustainable construction procurement. A guide to delivering environmentally responsible projects*
C571, CIRIA, London (ISBN 978-0-86017-571-1)

ALUMINIUM 73 (1997)
"Aluminium bridges and bridge decks for lowest life cycle costs"
*J. Aluminium*, vol 73, **II**, January, pp 777–781

ANGUS, H T (1976)
*Cast iron: Physical and engineering properties (2nd edition)*
Revised 1978, British Cast Iron Research Association, Butterworth & Co. Ltd, London,
(ISBN 0-40870-933-2)

ANON (2006)
"Bashed Battersea Bridge reopens early"
*New Civil Engineer,* 19 January

BAYLISS, D and DEACON, D (2002)
*Steelwork corrosion control*
Spon press Ltd, London (ISBN 978-0-41526-101-2)

BCSA (2005)
*Guide to the erection of steel bridges*
BCSA Publication no 38/05, British Constructional Steelwork Association Limited
(ISBN 0-85073-046-5) Available from: <http://www.steelconstruction.org>

BEALES, C and DALY, A F (1991)
"Weighing up the results"
*Surveyor*, vol 186, no 5167, pp 15-16

BERRIDGE, P S A (1969)
*The girder bridge after Brunel and others*
Robert Maxwell, UK (ISBN 978-0-08007-095-7)

BISHOP, R R (1986)
*Enclosure – an alternative to bridge painting*
Research Report RR 83, TRRL, Crowthorne

BRE (1985)
*Guide to practise in corrosion control. Corrosion of metals by wood*
Digest 301, Building Research Establishment, Garston (ISBN 851253555)
<http://www.brebookshop.com/>

BRIDGE DESIGN and ENGINEERING (1997)
"The light touch"
*Bridge design and engineering*, no 8, August, pp 57–59

BUSSELL, M (1984)
"Armstrong Bridge"
*Trust,* Tyne and Wear Industrial Monuments Trust, Nov 1984, pp 13–17

BUSSELL, M (1997)
*Appraisal of existing iron and steel structures*
SCI publication 138, Steel Construction Institute, Berks, UK (ISBN 978-1-85942-009-6)

CADEI, J M C, STRATFORD, T J, HOLLAWAY, L C and DUCKETT, W G (2004)
*Strengthening metallic structures using externally bonded fibre-reinforced polymers*
C595, CIRIA, London (ISBN 978-0-86017-595-7)

CHETTOE, C S, DAVEY, N and MITCHELL, G R (1944)
"The strength of cast iron bridges"
*Journal of the Institution of Civil Engineers*, no 8

CIVIL ENGINEERING (1996)
"New aluminium decks cut loads, add life"
*Civil Engineering*, ASCE, Vol 66, **8**, August, p 12

CLARK, K (2001)
*Informed conservation*
English Heritage, Swindon (ISBN 978-1-87359-264-9)

CLUBLEY, S K, WINTER, S N, AND TURNER, K W (2006)
Heat straightening repairs to a steel road bridge
*Bridge Engineering*, Institution of Civil Engineers, **159**, March 2006, pp 35–42

CORUS (2004)
*Weathering steel – connecting with the environment*
Corus Construction and Industrial, North Lincolnshire. Available from:
<http://www.corusgroup.com/file_source/StaticFiles/Business%20Units/CC&I/Products/Plates/Weathering_Steel.pdf>

CORUS/STEEL CONSTRUCTION INSTITUTE/BCSA (2003)
*Achieving sustainable construction – guidance for clients and their professional advisors*
Available from: <http://www.steel-sci.org/NR/rdonlyres/7411C850-6B64-45C6-A7F1-9F95C15F4C23/2383/AchievingSustainableConstruction.pdf>

COVENTRY, S, WOOLVERIDGE, C and HILLIER, S (1999)
*The reclaimed and recycled construction materials handbook*
C513, CIRIA, London (ISBN 978-0-86017-513-1)

COVENTRY, S and WOOLVERIDGE, C (1999)
*Environmental good practice on site*
C502, CIRIA, London (ISBN 978-0-86017-502-5)

*Design manual for roads and bridges*, The Stationery Office, London (2007)

Volume 1   BA 57/01 *Highways structures: approval procedures and general design*
           Section 3 Part 7 Design for durability (DMRB 1.3.7)

Volume 2   BD 7/01 *Weathering steel for highway structures*
           Section 3 Part 8 (DMRB 2.3.8)

Volume 3   BD 89/03 *The conservation of highway structures*
           Section 2 Part 4 (DMRB 3.2.4)

           BD 50/92 *Technical, requirements for the assessment and strengthening programme for highway structures: Stage 3 long span bridges*
           Section 4 Part 2 (DMRB 3.4.2)

           BD 21/01 *The assessment of highway bridges and structures*
           Section 4, Part 3 (DMRB 3.4.3)

           BA 16/97 *The assessment of highway bridges and structures*
           Section 4, Part 4 (DMRB 3.4.4)

           BD 63/07 *Inspection of highway structures*
           Section 4 Part 4 (DMRB 3.4.4)

BD 56/96 *The assessment of steel highway bridges and structures*
Section 4, Part 11 (DMRB 3.4.11)

BA 56/96 *The assessment of steel highway bridges and structures*
Section 4, Part 12 (DMRB 3.4.12)

BD 61/96 *The assessment of composite highway bridges and structures*
Section 4, Part 16 (DMRB 3.4.16)

BA 61/96 *The assessment of composite highway bridges and structures*
Section 4, Part 17 (DMRB 3.4.17)

BD 79/06 *Management of substandard highway structures*
Section 4 Part 18 (DMRB 3.4.18)

Volume 10   HA 80/99 *Environmental design and management: nature conservation, including advice notes on biodiversity and species-specific guidance. Nature conservation in relation to bats*
Section 4 Part 3 (DMRB 10.4.3)

DAWSON, S (2001)
"Walking through hoops"
*The Architects Journal*, 6–13 December, pp 41–42

DEACON, D H, ILES, D C, and TAYLOR, A J (1998)
*Durability of steel bridges – A survey of the performance of protective coatings*
Technical Report, Steel Construction Institute, Berks (ISBN 1-85942-081-8)

DE VOY, J and WILLIAMS, J M (2007)
*Strengthening Coalport Bridge*
International Association for Bridge and Structural Engineering, Zurich
<http://www.iabse.org/journalsei/asareader/vol17_2/Strengthening.php>

EUROPEAN STEEL DESIGN EDUCATION PROGRAMME (1995)
*Improvement techniques in welded joints*
(ESDEP) Lecture 12.5

ENVIRONMENT HYGIENE SERVICES (2007)
*Apparatus for preventing birds accessing a habitable part of a structure*
PCT Patent Application no PCT/GB2005/001426
<http://www.glideholdings.com/>

FAIRBAIRN, W (1864)
"Experiments to determine the effect of impact, vibratory action and long continued changes of load on wrought iron girders"
*Royal Society of London*, vol 13, 1863–1864, pp 121–126

FIRTH, I P T (1993)
"A tale of two bridges: the Lockmeadow and Halgaver Bridges"
*The Structural Engineer*, 5 March 2002, pp 26–31

GANNETT FLEMING (2003)
*Report on the July 21st collapse of the Kinzua Viaduct McKean County, Pa*
Pennsylvania Department of Conservation and Natural Resources, December 2003
<http://www.dcnr.state.pa.us/info/kinzuabridgereport/kinzua.html>

GILL, J, JAYASUNDARA, J, and COCKSEDGE, C (1994)
"Avonmouth Bridge – Assessment and strengthening"
In: *Proc Bridge modification conference*, ICE London, Thomas Telford, ASCE
(ISBN 978-072772-028-3)

GONSALVES, B F and DEACON, R W (1990)
"Docklands Light Railway and subsequent upgrading: general contract and design principles"
*Civil Engineering*, Institution of Civil Engineers, Part 1 1990, **88**, Aug, pp 601–617

GURNEY, T R (1991)
*Fatigue of welded structures 2nd edition*
Cambridge University Press, Cambridge (ISBN 0-52122-558-2)

GURNEY, T R (1992)
*Fatigue of steel bridge decks*
TRL State of the art review 8, TRL Crowthorne

HARDING, J E H, PARKE, G A R and RYALL, M J (eds) (1990)
"Bridge Management"
In: *Proc 1st Int conf on Bridge management*, Guildford, Elsevier Applied Science
(ISBN 1-85166-456-4)

HARDING, J E H, PARKE, G A R and RYALL, M J, (eds) (1993)
"Bridge Management"
In: *Proc 2nd Int conf on Bridge management*, Guildford, Thomas Telford
(ISBN 0-7277-1926-2)

HARDING, J E, PARKE, G A R and RYALL, M J (eds) (1996)
"Bridge management"
In: *Proc 3rd Int conf on Bridge management*, Guildford, E & F Spon, London
(ISBN 0-419-21210-8)

HARDING, J E, PARKE, G A R and RYALL, M J (eds) (2000)
"Bridge management"
In: *Proc 4th Int conf on Bridge management*, Guildford, Thomas Telford
(ISBN 0-7277-2854-7)

HAYWARD, A C G, SADLER, N and TORDOFF, D (2002)
*Steel bridges – a practical approach to design for efficient fabrication and construction*
BCSA Publication no 34/02, British Constructional Steelwork Association Limited, London (ISBN 0-85073-046-5)

HEALTH AND SAFETY EXECUTIVE (2005a)
*The work at height regulations 2005, a brief guide*
INDG401(rev1), HSE Books, Sudbury (ISBN 978-0-71766-231-9)
<http://www.hse.gov.uk/pubns/indg401.pdf>

HEALTH AND SAFETY EXECUTIVE (2005b)
*Safe work in confined spaces*
INDG258, HSE Books, Sudbury (ISBN 0-71761-442-5)
<http://www.hse.gov.uk/pubns/indg258.pdf>

HILMAN ROLLERS
Information available from: <http://www.hilmanrollers.com/>

ILES, D C (2004)
*Design guide for steel railway bridges*
Steel Construction Institute, Berks (ISBN 1-85942-150-4)

INSTITUTION OF CIVIL ENGINEERS (1998)
*Guidelines for the supplementary load testing of bridges*
Thomas Telford Ltd, London (ISBN 978-0-72772-737-4)

JACKSON, P A (1996)
"The analysis and assessment of bridges with minimal transverse reinforcement"
In: *Proc 3rd Int conf on Bridge management,* University of Surrey, Guildford, 14–17 April,
pp 779–785

JACKSON, P A (2001)
"Is bridge assessment losing its credibility? Viewpoint"
*The Structural Engineer,* vol 79, **9,** May 2001, pp 15–16

JAMES, J G (1981)
"The evolution of iron bridge trusses to 1850"
Reprinted 1997, *The Newcomen Society,* vol 52, pp 67–101 Sutherland, R J M (ed)
*Structural iron, 1750-1850: studies in the history of civil engineering,* vol 9, pp 311–345,
Ashgate Variorum. Available from: <http://www.pubs-newcomen.com>

KAUFMANN, E and CONNOR, R (2003)
*Evaluation of the anchor bolt components. Kinzua bridge collapse*
ATLSS Engineering Research Centre, Lehigh University
<http://www.dcnr.state.pa.us/info/kinzuabridgereport/app/appc.pdf>

KENNEDY-REID, I L, MILNE D M and CRAIG, R E (2001)
*Steel bridge strengthening: a study of assessment and strengthening experience and identification
of solutions*
Thomas Telford, London (ISBN 978-0-72772-881-4)

LALLYETT, W (2001)
"Lightweight aluminium bridges the gap at Wimbledon"
*The Architects Journal,* 6–13 December, pp 125–126

LARSSON, T (2006)
*Material and fatigue properties of old metal bridges*
Licentiate Thesis, University of Lulea, Sweden
<http://epubl.ltu.se/1402-1757/2006/26/LTU-LIC-0626-SE.pdf>

LEE, D J (1990)
*Bearings and expansion joint 2nd edition*
Taylor and Francis Group, London (ISBN 978-0-41914-570-7)

LUL (2006)
*Engineering standard 2-01304-002, civil engineering– bridge structures*
London Underground Limited, London

McKIBBINS, L D, MELBOURNE, C, SAWAR, N and GALLIARD, C S (2006)
*Masonry arch bridges: condition appraisal and remedial treatment*
CIRIA C656, London (ISBN 978-0-86017-656-5)

McLESTER, R (1988)
"Railway component fatigue testing"
*Full-scale fatigue testing,* Marsh, K J (ed), Butterworth-Heinemann
(ISBN 978-0-40802-244-6)

MADDOX, S J (1991)
*Fatigue strength of welded structures, Cambridge, 3rd edition*
Woodhead Publishing, Gresham Books, Abington (ISBN 978-0-84931-774-3)

MATTHEWS, S J and OGLE, M H (1996a)
"Investigation and load testing of a steel lattice truss viaduct"
*Bridge Management 3,* Harding, J E, Parke, G A R, and Ryall, M J (eds), University of
Surrey, E&F N Spon, London

MATTHEWS, S J and OGLE, M H (1996b)
"Refurbishment of a steel lattice truss viaduct"
*Bridge Management 3*, Harding, J E, Parke, G A R, and Ryall, M J (ed), University of
Surrey, E&F N Spon, London

MEHRKAR-ASL, S, BROOKES, C L and DUCKETT, W (2005)
"Saving half through girder bridges using non-linear finite element analysis"
In: *Proc 5th Int conf Bridge management,* University of Surrey, Guildford, 11–13 April

MERRISON (1974)
"Inquiry into the basis of design and erection of steel box girder bridges"
*Report of Committee. Appendix I Interim design and workmanship rules*, HMSO, London

METAL STITCHING PROCESSES
Details of process available from (among others):
<http://www.metalock.co.uk/>, <http://www.locknstitch.com/Metal_Stitching.htm>,
<http://www.in-situ.co.uk/metalstitching.html>

MILLER, G (2006)
"Union chain bridge: linking engineering"
*Civil Engineering,* Institution of Civil Engineers, vol 159, **2**, May pp 88–95

MINISTRY OF TRANSPORT (1970)
*The assessment of highway bridges for construction and use vehicles*
Technical memorandum (bridges) BE4/67 (including amendments up to 1970)

MORLEY, A (1947)
*Strength of materials*
No 011255, Longman, London. Available from:
<http://ukbookworld.com/cgi-bin/order_enq.pl?add=brockwells%23011255>

MULLINS, L (2007)
*Management and Organisational Behaviour*
Longmans-Pearson Education Limited, Essex (ISBN 978-1-40585-477-1)

MYLIUS, A (2005)
"Sandwich course"
*New Civil Engineer*, 18–15 August, pp 29–30

NARAYANAN, R (ed) (1991)
*Structures subjected to repeated loading*
Spon Press Ltd, London (ISBN 978-1-85166-567-9)

NATIONAL PHYSICAL LABORATORY (2004)
*Basics of corrosion-bimetallic corrosion*
<http://www.npl.co.uk/server.php?show=nav.1017>

NETWORK RAIL (2000)
*The use of BD and BA 61 for cased and filler beam bridges*
(unpublished but available for assessment of Network Rail structures)
Current Information Sheet 23, Network Rail, London

NETWORK RAIL (2002)
*Protective treatment for railway infrastructure*
RT/CE/S/039 Specification RT98, Network Rail, London

NETWORK RAIL (2002)
*Application and reapplication of protective treatment to railway infrastructure*
RT/CE/C/002, Network Rail, London

NETWORK RAIL (2004a)
*The structural assessment of underbridges*
RT/CE/C/025 (due to be renumbered as NR/SP/Civ/25), Network Rail, London

NETWORK RAIL (2004b)
*Examination of structures*
NR/SP/CIV/017 (formerly RT/CE/S/017), Network Rail, London

NETWORK RAIL (2005)
*Standards of competence for examination of structures*
NR/SP/CIV/047, Network Rail, London

NEW CIVIL ENGINEER (1997)
"Design blamed for Israeli bridge fall"
*New Civil Engineer*, 17 July, p 3

NEWTON, J, WILLIAMS, C, NICHOLSON, B, VENABLES, R, WILLETTS, R and
MOSER, B (2004)
*Working with wildlife. A resource and training pack for the construction industry*
CIRIA C587, London (ISBN 978-0-86017-587-2)

OWENS, G W and KNOWLES, P R (ed) (2003)
*Steel designers manual, 6th edition*
Steel Construction Institute, Berks (ISBN 978-0-63204-925-7)

PARKE, G A R AND DISNEY, P (ed) (2005)
"Bridge management"
In: *Proc 5th Int conf on Bridge management,*Guildford, 11–13 April 2005, **5**, Thomas
Telford, London (ISBN 978-0-72773-354-2)

PERRY J, PEDLEY M, and BRADY, K (2003a)
*Infrastructure embankments – condition appraisal and remedial treatment. 2nd edition*
CIRIA C591, London (ISBN 978-0-86017-591-9)

PERRY, J, PEDLEY, M, and REID, M (2003b)
*Infrastructure cuttings – condition appraisal and remedial treatment*
CIRIA C592, London (ISBN 978-0-86017-592-6)

PHARES, B, ROLANDER, D, GRAYBEAL, B, MOORE, M and WASHER, G (2001)
*Reliability of visual inspection of highway bridges*
Report FHWA-RD-01-020, Federal Highway Administration, Washington DC
<http://www.tfhrc.gov/hnr20/nde/01105.pdf>

PILGRIM, D and PRITCHARD, B P (1990)
"Docklands Light Railway and subsequent upgrading: design and construction of
bridges and viaducts"
*Civil Engineering*, Institution of Civil Engineers, **1**, Aug, pp 619–638

PRITCHARD, B (1992)
*Bridge design for economy and durability: Concepts for new, strengthened and replacement bridges*
Thomas Telford, London (ISBN 978-0-72771-671-2)

REES (2002)
*Urban environments and wildlife law, a manual for sustainable development*
Blackwell Publishing, Oxford (ISBN 978-0-63205-743-6)

RESTORATION (1983)
"Spare cash saves otiose Newcastle bridge"
*New Civil Engineer*, 31 March

SCI STEEL BRIDGE GROUP (2002)
*Guidance notes on best practice in steel bridge construction*
Publication no SCI-P185, Steel Construction Institute, Berks

SIMON, P, HILLSON, D, and NEWLAND, K (eds) (1997)
*Project risk analysis and management guide*
APM Group Limited, High Wycombe (ISBN 978-0-95315-900-0)

SREEVES, J (2007)
"Future-proof: Upton upon Severn viaduct, United Kingdom"
*Civil Engineering*, Institution of Civil Engineers, 160, pp 33–38

STEELE, K (2004)
*Environmental sustainability in bridge management*
IP14/04, Building Research Establishment, Garston, (ISBN 978-1-86081-732-8)

STREETEN, A D F (1990)
"How to handle scheduled and listed structures"
In: *Institution of highways and transportation national workshop on concepts for the management of highway structures*, March 1990, pp 29–44

SWAILES, T (2006)
*Scottish iron structures*
Guide for Practitioners 5, Historic Scotland, Edinburgh (ISBN 1-90496-612-8)

THE STRUCTURAL ENGINEER (1997)
"Bridge decking"
*The Structural Engineer*, vol 75, **8**, 15 April, p A10

TILLY, G P (2002)
*Conservation of bridges: a guide to good practice*
Spon Press, London (ISBN 978-0-41925-910-7)

TRINIDAD, A A (1993)
"Aluminium highway bridges in the USA"
In: *Proc 2nd Int conf on Bridge management*, Harding, J E *et al* (ed), pp 190–199, Thomas Telford, London

TWELVETREES, W N (1900)
*Structural iron and steel*
Fourdrinier

TYLER, M and LAMONT, D R (2005)
"Construction health and safety"
*Construction Law Handbook*, Ch 3.2, Thomas Telford, London (ISBN 978-0-72773-485-3

FEDERAL HIGHWAY ADMINISTRATION (1998)
*Heat straightening repairs of damaged steel bridges – a technical guide and manual of practice*
FHWA-IF-99-004, US DoT. Available from <http://www.fhwa.dot.gov/bridge/heat.htm>

UK BRIDGES BOARD (2005)
*Management of highway structures – a code of practice*
The Stationery Office, London (ISBN 978-0-11552-642-8)
<http://www.ukroadsliaisongroup.org/pdfs/p02_management_of_highway_structures.pdf>

WEBSTER, P and MEHRKAR-ASL, S (2002)
"Battersea yard – bridge infill project"
*Railway Engineering News 2000*
<http://www.mehrkar.com/shapour/PDFs/BATTER.PDF>

XIE, M, BESSANT, G, CHAPMAN, J C and HOBBS, R E (2001)
"Fatigue of riveted bridge girders"
*The Structural Engineer*, vol 79, **9**, May, pp 27–36

## Acts and Directives

HER MAJESTY'S STATIONERY OFFICE (1968)
Road Transport Act 1968, Section 117

HER MAJESTY'S STATIONERY OFFICE (1972)
Railway Bridges (Load Bearing Standards) Regulations (England and Wales) order 1972 (SI 1705/1972)

The Health and Safety at Work Act of 1974

Ancient Monuments and Archaeological Areas Act 1979

Planning (Listed Buildings and Conservation Areas) Act 1990

New Roads and Street Works Act (NRSWA) 1991

European Marketing and Use Directive (89/677/EEC)

## Regulations

Environmental Protection (Controls on Injurious Substances) Regulations 1992

Environmental Protection (Duty of Care) Regulations 1992

Construction Design and Management Regulations (CDM) 1994 (as amended 2007)

Construction (Health, Safety and Welfare) Regulations of 1996 (under revision, 2007)

Confined Spaces Regulations 1997

Lifting Operations and Lifting Equipment Regulations (LOLER) 1998

Provision and Use of Work Equipment Regulations 1998

Management of Health and Safety at Work Regulations of 1999

Control of Lead at Work (CLAW) Regulations 2002

Control of Substances Hazardous to Health (COSHH) Regulations 2002

Working at Height Regulations (WAHR) 2005

Noise at Work Regulations 2005

## Standards

BS 153-3 and 4:1968 *Specification for steel girder bridges; part 3b stresses and part 4 design and construction* (reset and reprinted April 1966 and including subsequent amendments up to Amendment no. 8 September 1968)

BS 5400-10:1980 *Steel, concrete and composite bridges. Code of practice for fatigue*

BS 7079:1989 *Visual surface preparation standards of steel prior to painting* (an appendix to the ISO 8501 Part 1)

BS 5400-6:1999 *Steel, concrete and composite bridges. Specification for materials and workmanship, steel*

BS 7985:2002 *Code of practice for the use of rope access methods for industrial purposes*

BS EN 10045-1:1990 *Charpy impact test on metallic materials. Test method (V- and U-notches)*

BS EN 24506:1993 *Specification for hard metals. Compression test*

BS EN 1712:1997 *Non-destructive testing of welds. Ultrasonic testing of welded joints acceptance levels*

BS EN 571-1:1997 *Non-destructive testing. Penetrant testing. General principles*

BS EN 1290:1998a *Non-destructive testing of welds - magnetic particle testing of welds*

BS EN 1713: 1998b *Non-destructive examination of welds. Ultrasonic examination*

BS 5400-3:2000 *Steel, concrete and composite bridges. Code of practice for the design of steel bridges*

BS EN 1711:2000 *Non-destructive examination of welds - eddy current examination of welds by complex plane analysis*

BS EN 10002-1:2001 *Tensile testing of metallic materials. Method of test at ambient temperature*

BS EN ISO 8501-1:2001 *Preparation of steel substrates*

BS EN 10025-1:2004 *Hot rolled products of non-alloy structural steels. General delivery conditions*

BS EN 1993-1-8:2005 *Eurocode 3 design of steel structures. Part 1-8 design of joints*

BS EN 1993-1-9:2005 *Eurocode 3 design of steel structures. Part 1-9 fatigue*

BS EN 10164:2004 *Steel products with improved deformation properties perpendicular to the surface of the product – technical delivery conditions*

BS 7910:2005 *Guide to methods for assessing the acceptability of flaws in metallic structures*

ISO 8501-1:2001 (updated 2007) *Preparation of steel substrates before application of paints and related products — Visual assessment of surface cleanliness. Part 1: Rust grades and preparation grades of uncoated steel substrates and of steel substrates after overall removal of previous coatings*

PD 6484:1979 *Commentary on corrosion at bimetallic contacts and its alleviation*

# A1 Aluminium bridges

> This appendix describes the development of aluminium alloy bridges giving examples and performances. The pros and cons of using aluminium alloys in new construction and for strengthening are summarised.

## A1.1 INTRODUCTION

Aluminium alloys have been of interest to bridge engineers since the 1930s, as they have some attractive mechanical properties and are seen as requiring a low level of maintenance requirements. Moreover the material has some strong advocates led by the aluminium manufacturers.

In countries using de-icing salt during winter maintenance and experiencing corrosion in bridge decks and beams, materials having good corrosion resistance have obvious potential. Nevertheless, despite an upturn in interest in recent years, aluminium bridges are comparatively rare and remain behind timber in numbers and popularity.

## A1.2 EXAMPLES OF ALUMINIUM BRIDGES

There are about 10 aluminium highway bridges in the United States, the first being a 30.5 m span rail bridge across Grasse River, New York, built in 1946. Between 1958 and 1963, when steel was in short supply, five more aluminium bridges were built. For some of the bridges a shop fabricated multi-voided extruded deck panel, shown in Figure A1.1 was used for some of the bridges. At about 123 kg/m² a 3 m × 12 m panel weighs about 4.5 tons. The system enables rapid construction and is an attractive option for replacement of deteriorated concrete decks (Bridge Design and Engineering, 1997).

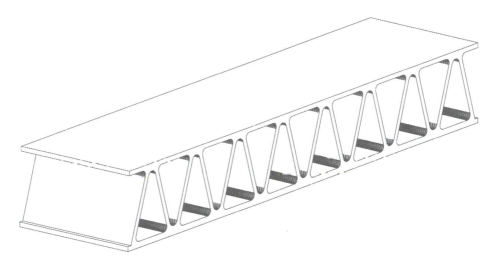

**Figure A1.1**     *Cross-section of Reynolds' Alumadeck*

The first long-span aluminium bridge was the 88.4 m span highway bridge with a riveted box arch, across Saguenay River in Canada, built in 1950 as described by Trinidad (1993).

In the last 15 years there has been research and construction in the Nordic countries where there are now about 40 aluminium highway bridges, mainly in Sweden.

In Norway an initiative supported by the Norwegian Research Council led to the construction of Forsmo Bridge, a two-span 39 m highway bridge built in 1995 (Aluminium 73, 1997). The structure is composed of welded box girders with an isotropic deck plate fabricated in a single section weighing 28 tons, see Figure A1.2. Speed of construction and installation (a few hours), low maintenance costs and light weight were listed as reasons for using aluminium for this bridge.

Also in Norway, studies have been carried out for a 300 m highway bridge in a harbour and industrial environment. Cost estimates at the time of the studies showed that aluminium had the highest initial cost:

- aluminium alloy, $7.4m
- steel, $6m
- concrete $5.5m.

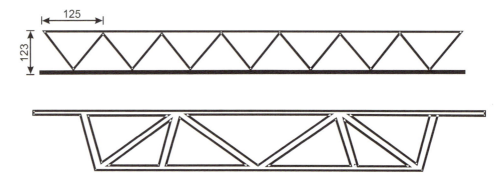

**Figure A1.2**    ***Typical sections of Forsmo Bridge, Norway***

In Britain several aluminium bridges have been constructed over the years:

- a twin-bascule aluminium bridge constructed in Sunderland docks in 1948 see Figure A1.3
- Pitlochry footbridge, built c1975
- a canal lift bridge in Gloucester, c1975
- enclosure of the steel beams of Conon Bridge, a three-span (34-44-34 m) river bridge near Inverness (1982)
- West India Quay (The Structural Engineer, 1997), a floating footbridge in Docklands, London, having a total length of 90 m, 1996
- Princes Dock footbridge, Liverpool, an architectural structure of 30 m span built in 2001, see Figure A1.4 (Dawson, 2001)
- Lockmeadow (Firth, 1993), a two-span cable-stayed footbridge having a main span of 46 m across the River Medway (2001)
- Bailey type portable footbridges (Lallyett, 2001).

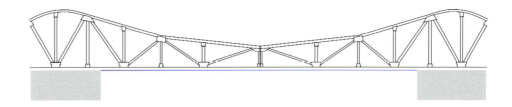

**Figure A1.3**        *Twin bascule aluminium bridge, Sunderland*

**Figure A1.4**        *Princes Dock Bridge, Liverpool*

# A1.3        ENCLOSURE

Enclosure is a method of corrosion protection developed in the 1980s and supported by theoretical studies and fieldwork (Bishop, 1986). Two pilot enclosure schemes were carried out successfully in West Sussex followed by New Conon Bridge which was the first full scale scheme, see Figure A1.5. Costs of construction in fibreglass and aluminium alloy were obtained and found to be very close. It was decided to select aluminium alloy anodised in a colour similar to the beams being protected. Subsequent enclosures carried out by other organisations have been in fibreglass and although aluminium alloy has not been seriously considered again, it remains a competitive material for enclosures. It is necessary to be aware of the possible risk of bi-metallic corrosion and to ensure that there is no direct contact between the dissimilar metals.

**Figure A1.5**    *Enclosure of New Conon Bridge, Scotland*

## A1.4    RETROFITTED DECKS

The earliest deck to be replaced with aluminium was Smithfield Street Bridge in Pittsburgh, Pennsylvania, built in 1882 and re-decked with an aluminium deck plate and stringers in 1933. This deck was replaced in 1967 with a lighter weight orthotropic aluminium deck enabling the live load capacity to be increased further (see Figure A1.6).

The deck of an obsolete bridge of 17 m span in Virginia was replaced with aluminium panels to increase the width from 7–8.5 m and reduce dead load by some 35 per cent. The cost was estimated to be somewhat higher than conventional concrete (Civil Engineering, 1996).

Corbin Bridge, Pennsylvania, is a 91 m heritage-listed steel suspension bridge. Its aging steel deck was replaced with aluminium panels to reduce weight by 50 per cent and raise permissible vehicular loads from a posting of seven tons to allowing 24 ton vehicles, one at a time.

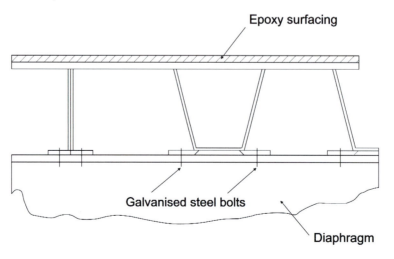

**Figure A1.6**    *Cross-section of orthotropic deck of Smithfield Bridge, Pennsylvania*

## A1.5  MATERIAL PROPERTIES

### The alloys

Aluminium alloys have a range of properties designed to meet different applications. It is important to select the right alloy to suit the method of fabrication and specific application, eg strain hardened alloys for structures with welded connections.

The physical properties of a typical aluminium alloy are summarised in Table A1.1 in relation to those of steel.

Table A1.1    *Properties of a typical aluminium alloys and steel*

| Property | Aluminium | Steel |
|---|---|---|
| Melting temperature | 660°C | 1500°C |
| Density | 2 710 kg/m³ | 7850 kg/m³ |
| Elastic Modulus | 70 kN/mm² | 205 kN/mm² |
| Coefficient of thermal expansion | $23 \times 10^{-6}$ per°C | $12 \times 10^{-6}$ per°C |

In relation to fatigue performance, the reference stress ranges at 2 m cycles for the design of aluminium joints are from 14 N/mm² to 60 N/mm².

### Advantages of aluminium

The main advantages of using aluminium alloys in the construction, enhancement and renewal of bridges can be summarised as follows:

- the light weight is advantageous to the transport of fabricated sections to site and their erection. It has obvious attractions for use in temporary (Bailey type) and movable bridges

- the good resistance to corrosion of correctly selected alloys has advantages in the management and maintenance of bridges in climates where de-icing salts are used, and in a marine environment

- the reduced maintenance requirements can lead to whole life costs that offset the higher initial costs

- the material has an attractive appearance and can be used in architectural footbridges.

### Disadvantages of aluminium

The advantages have to be set against various disadvantages, some of which can be avoided by careful design, others which have to be accepted:

- aluminium alloy sections are more expensive than either steel or concrete, albeit the differentials are influenced by contemporary world metal prices

- although corrosion resistance can be good, special care has to be taken to select the correct alloy

- bi-metallic corrosion can occur in composite structures where aluminium is close to or in contact with steel

- impact resistance is poor. Notches can be created which form potential initiation points for fatigue, and sections can be buckled under bridge bashing and other impacts

- resistance to fire is poor due to the low softening and melting temperatures

- it is difficult to obtain good adhesion between wearing courses and aluminium deck plates

- the mode of collapse of an Israeli aluminium footbridge is indicative that special care should be taken in ULS design (New Civil Engineer, 1997).

## A1.6 PERFORMANCE IN PRACTICE

Most of the information about performances in practice is obtained from experiences in the United States where there are several bridges that are now about 30 years old. Their performances have been considered to be satisfactory, albeit there have been reports of fatigue cracking and minor corrosion (Trinidad, 1993). The 1933 aluminium deck panels on Smithfield Street Bridge developed corrosion and were replaced in 1967 by an orthotropic deck constructed of a more corrosion resistant alloy. When problems with the wearing surface prompted another upgrade in 1994, the opportunity was taken to assess the condition of 110 m² of the worst sections. It was concluded that while there were wearing surface problems, the aluminium alloys had generally performed well.

In Israel, a simply-supported tubular aluminium footbridge of 20 m span collapsed in 1997. The bridge was reported to have folded in mid-span under crowd loading. It was decided that it had been under-designed (New Civil Engineer, 1997).

In Britain experience has been rather mixed with occurrences of corrosion and fatigue cracking.

**Figure A1.7**     *Welded repair to fatigue crack in aluminium structure*

The bascule bridge in Sunderland docks was in a marine environment and developed excessive corrosion to the extent that it was deemed necessary to be demolished after 20 years. It was suggested that an alloy with inadequate resistance to corrosion had been selected, and it was also suggested that bi-metallic corrosion had occurred.

After 30 years in service the lift bridge in Gloucester had cracked at the rivet holes, a classic location with stress concentrations to encourage the initiation of fatigue. The crack had been repaired by welding shown in Figure A1.7, but not surprisingly, the cracking recurred through the weld and a second repair was made by bolting a plate over the area.

The anodised aluminium enclosure of the main beams of New Conon Bridge has been in service for about 20 years, and when last inspected was found to be in pristine condition, and fully successful in protecting the steel beams from corrosion. In contrast, corrosion has developed in steel fittings outside the enclosure and in the aluminium parapet posts. There was evidence of guano droppings from birds that had entered but were unable to escape from inside the enclosure, and it remains unknown whether any corrosion will result.

Bearing in mind that good fatigue and corrosion properties are a major attraction of aluminium alloys, it is surprising that there have been occurrences of these problems in such a small population of bridges. This may be due to designers being over confident, and failing to make adequate allowances for fatigue and corrosion.

## A1.7     POTENTIAL FOR ALUMINIUM

Unless the relative costs of steel and aluminium change, wholly aluminium bridges are unlikely to become competitive with steel and concrete. The potential for aluminium alloys in bridges is:

- architectural footbridges where the appearance and light weight of aluminium can be used to advantage

- components of portable (Bailey type) and movable bridges where light weight is all-important

- retrospective replacement of deteriorated concrete decks, an application that is attractive in the US where there is a burgeoning market. It is of less interest in Britain where waterproof membranes are providing good protection, and requirements for replacement decks are less common

- ancillaries such as enclosures to protect steelwork from weathering.

Care is required in selecting the correct alloy for the application and method of construction. It should not be assumed that aluminium alloys will not corrode or suffer fatigue damage.

# A2 Case studies

This appendix provides 11 case studies of repair and strengthening works on iron and steel bridges exemplifying a range of situations and techniques. The case studies have been selected to provide a representative range of methods and examples including schemes that were unsuccessful, temporary, and reconstruction, as well as repairs designed, to be permanent.

### A2.1 Coalport Bridge*, Shropshire

Key words: strengthening, load testing, monitoring, bonded plating, lightweight deck.

### A2.2 Battersea Bridge, London

Key words: ship impact, crack detection, repair, bonded plating, metal stitching.

### A2.3 Kinzua Viaduct*, Pennsylvania, USA

Key words: repair, hidden defects, collapse.

### A2.4 Braemore Great Bridge*, Hampshire

Key words: corrosion, strengthening, modified structural action, replacement deck.

### A2.5 Brockhampton Bridge, Hampshire

Key words: vehicular impact, repair, heat straightening.

### A2.6 Battersea Yard Railway Bridge, South London (courtesy TFL)

Key words: strengthening, infilling, monitoring.

### A2.7 Pontcysyllte Aqueduct*, North Wales

Key words: corrosion, repainting.

### A2.8 Armstrong Bridge*, Newcastle

Key words: corrosion, repair, changed articulation.

### A2.9 Balgay Park Footbridge, Dundee

Key words: failed repair, added concrete, corrosion.

### A2.10 Killearn Station Bridge, Stirlingshire

Key words: corrosion, reconstruction.

### A2.11 Holloway Road Bridge, London

Key words: corrosion, temporary repair.

Note  * Bridge recognised nationally as having special historic significance.

Other case studies, not given here, can be found in publications such as Kennedy-Reid *et al* (2001), where 39 examples are described. In Cadei *et al* (2004) descriptions of three bridge schemes plus references to 10 others are given, and Tilly (2002) deals with historic structures. Pritchard (1992) cites various examples of strengthening in relation to achieving minimal disruption to traffic. There are also some 22 cases referenced in the Bibliography.

# A2.1    COALPORT BRIDGE, SHROPSHIRE

> Strengthening of an early cast iron bridge located in a world heritage site.

The historic Coalport Bridge carries the class 3 Broseley to Brockton road over the river Severn in the village of Coalport. It is owned jointly by Shropshire County Council (SCC) and the Borough of Telford and Wrekin (BTW). The bridge is a Grade II* listed building and a scheduled ancient monument, and is situated within the Ironbridge Gorge World Heritage Site. Prior to its strengthening the bridge was signed with a two ton weight limit and crossings were limited to one vehicle at a time.

The present bridge was built between 1799 and 1800 with a cast iron arched structure supporting a timber deck. There were originally three cast iron ribs until 1818 when the bridge was widened with the addition of two further ribs, and the timber deck was replaced with the cast iron deck that is still present today.

**Figure A2.1**    **Coalport Bridge during strengthening work**

This iron deck supported a road of cinders and cobbles which was later surfaced and eventually replaced in the early 1980s with a reinforced concrete deck slab. The iron deck plates are supported off cast iron inverted "T" beams (originally by wrought iron shims but latterly by wooden wedges) which are supported off the arch ribs by cast iron vertical supports of varying length depending on their position along the bridge.

The main river span is approximately 32 m between the arch rib springing points, and on either side of the river span there are short approach spans. In addition the south abutment is hollow and contains two more short spans making a total of five spans.

The bridge carries about 2000 vehicles a day. It was assessed by SCC as having zero live load capacity to BD 21/01 but rather than close the bridge to traffic the bridge engineer at the time took the view that the bridge should be able to carry an additional one per cent of its self weight as live load. It was estimated that the bridge weighed 200 tons so a two ton limit was arrived at, with only one vehicle at a time permitted on the bridge. The County experimented with various methods to enforce this restriction, without success, before erecting goalpost type height and width restrictions.

A view of the bridge during strengthening is shown in Figure A2.1.

## The problem

In 2001 SCC, BTW and English Heritage took steps to safeguard the future of the bridge, to improve the traffic calming and to strengthen the bridge with an aspiration to achieve 7.5 ton loading and remove the one vehicle at a time restriction which had proved impossible to police.

For a structure of this vintage, there were no original drawings available, and it became apparent that a complete dimensional and structural survey of the bridge would be required before the current load carrying capacity of the bridge could be assessed, much less develop a strengthening scheme.

When the survey was completed there was sufficient information to construct a sophisticated finite element model of the main river span. The side spans plus the two hidden spans in the south abutment, known locally as the undercroft, are accessible to simple calculation but the main river span is not.

Initially it was hoped that the analysis would show that strengthening was not required and that the bridge was stronger than initially believed, but this proved not to be the case and the analysis confirmed the original rating of dead load only.

## The solution

Different ways of strengthening the bridge were examined and it was concluded that the most practical was by plate bonding.

The behaviour of the bridge is very sensitive to the support conditions at the ends of the ribs and the cast iron inverted T-beams. Assumptions were made as to what these support conditions might be but when they were plugged into the model it was clear that either the existing cast iron was massively stronger than the allowable code values or the assumed supports were wrong.

In attempting to refine the estimates for these support conditions, an earlier survey when the movements of the bridge had been monitored for a year, was revisited. These data indicated that the bridge appeared to be moving an alarming amount. A later survey had found that the eastern vaults, of the undercroft were significantly distorted. The concern was that if these movements were correct then this could be the cause of the distortion observed in the vaults and that this movement would be likely to render redundant any strengthening scheme for the vaults. There was scepticism of the results of the movement survey but they could not be dismissed, so the bridge was periodically revisited and surveyed over a period of a year. Some small movements were measured but nothing like the range from the earlier survey.

One of the problems in developing the strengthening scheme is that the BD 21/01 code value for the permissible tensile stress of cast iron is not more than 46 N/mm². This is understood to be a conservative value but using a higher value requires justification and there were no spare parts of the bridge that could be cut up and tested. This problem was compounded by the fact that there are two bridges – the 1799/1800 bridge and the 1818 bridge, with two different cast irons. Early cast iron is an inherently brittle and variable material so to have any confidence in the result, it is necessary to test multiple samples, however there were no spare samples to test.

It was not possible to test the material but it was possible to test the support conditions. Over 200 vibrating-wire strain gauges were attached to the bridge and a data-logging system was designed to record its behaviour every 15 minutes over a two month period. During a temporary road closure an accurately weighed two ton vehicle was driven across to measure the response to live loading. The results from the monitoring and the live load test were then used to calibrate the finite element model.

## The strengthening scheme

The methodology for strengthening the main span was for the existing deck slab to be removed by hydrodemolition (water jetting) to avoid damaging the existing iron deck plates. Grade 355 steel was selected in preference to CFRP for the strengthening as the latter was found to be inappropriate in this instance as well as very expensive. Steel plates of 16 mm thickness were bonded to the central 6 m of the ribs, with the slab removed. The plates were clamped at their ends to resist peeling forces. A new structural lightweight reinforced concrete slab was then cast which spans between the verticals removing the need for the wooden wedges to support the deck. This procedure was followed in order to attract permanent load into the steel plates, leaving as much as possible of the permissible 46 N/mm² tensile capacity of the cast iron available for live load.

New connections were introduced at the top of the vertical supports to transfer load from the deck into the verticals. Steel plates with shear connectors were inserted between the deck plates above the verticals before the new deck was cast and these were clamped to a new bearing arrangement seated at the top of the web of the T-beams directly above the verticals, see Figure A2.2.

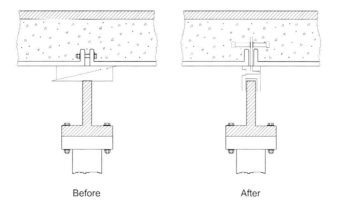

Before                  After

**Figure A2.2**       ***Tops of the vertical supports before and after strengthening***

It was not possible to meet the client's original aspiration for a 7.5 ton live load capacity, but it was raised to three tons and the one vehicle at a time restriction was removed.

Details of the work are given by De Voy and Williams (2007).

## A2.2  BATTERSEA BRIDGE, LONDON

> Repair of a river bridge having cast iron components impacted and damaged by an errant barge.

Battersea Bridge was constructed between 1886 and 1890 to a design by Sir Joseph Bazalgette. The bridge has five arch spans which are symmetrical about the main span and are 36.7 m, 47 m, 54.1 m, 47 m and 36.7 m respectively. The 12 m wide concrete deck is supported on inverted buckle plates spanning between longitudinal girders, which are in turn supported on spandrel columns, all formed from wrought iron. The spandrel columns are carried by seven segmental cast iron arch ribs in each span. Each rib was originally composed of five segments making a total of 175 cast iron segments.

On 20 September 2005 Battersea Bridge was struck by an errant gravel barge at the centre of the north intermediate span (span two). The barge was travelling in the downstream direction.

Rope access inspections carried out following the incident revealed damage to the central area of span two including the centre areas of the upstream and downstream edge cast iron ribs, the upstream coving units including supports, the upstream fascia panel and the upstream navigation lights and equipment. Some of the damage is shown in Figure A2.3.

**Figure A2.3**   *Damage caused by the impact*

The damage to the cast iron rib on the upstream side consisted of partial loss of section to the lower flange over an approximate 800 mm length and two cracks were detected using magnetic particle imaging, see Figure A2.4. One crack occurred at the location of the section loss to the flange and extended 245 mm through the remaining width of the flange and continued up through the web for an approximate length of 600 mm. The other crack was located 3 m further north and consists of a diagonal crack in the web of approximately 1900 mm length. The damage to the downstream rib consisted of section loss to the lower flange of the edge cast iron rib over an approximate 800 mm long section. Inspections revealed no apparent cracking in this area.

**Figure A2.4**   *Cracking shown by magnetic particle detection*

It is not clear how the damage was caused to the downstream rib although it is possible that the barge became lodged onto the lower flange immediately following the impact on the upstream side. As the tide was ebbing at the time of the incident this would then have resulted in the thrust from the barge being gradually transferred to the downstream rib which resulted in a local failure of the flange.

The damage was sufficiently serious that it was necessary for the bridge to be closed to motor vehicles except buses until repairs had been carried out. At one stage it was thought that the ribs were beyond repair. However, use of a computer model developed in support of strengthening work in the 1980s made it possible to rule out the necessity to replace the damaged arch ribs.

Repairs included bonding and bolting 20 mm thick mild steel plates to the underside of the lower flanges of the I-section ribs, see Figure A2.5.

**Figure A2.5**      *Bonded steel plating*

The cracks were stitched together with the Metalock system of nickel steel multi-dumbel connectors, as shown in Figure A2.6. See Section 9.3.4 for a description of metal stitching.

**Figure A2.6**      *Cracking repaired by metal stitching*

The main problem encountered was cold weather which could have affected the epoxy adhesive used to bond the plates. This was solved by using radiant heaters.

The repair work was started in December 2005 and was completed in time to re-open the bridge to traffic in early January 2006.

This case study is reproduced courtesy of Transport for London

# A2.3    KINZUA VIADUCT, PENNSYLVANIA, USA

> Collapse of a disused rail bridge during repair work.
> - fractures leading to the collapse had occurred in anchor-bolt assemblies at locations hidden from view
> - it had been presumed from external appearance that strengths of the anchor bolts were within commonly accepted values for wrought iron with reasonable account taken for corrosion losses
> - 75 per cent of the hold-down assemblies had pre-existing fatigue cracks.

Kinzua gorge lies in a high plateau 640 m above sea level in north central Pennsylvania. The first Kinzua Viaduct was constructed in 1882 for the New York, Lake Erie and Western Railroad and Coal Company. The 41 span 625 m long structure was constructed mainly of wrought iron. At the time of construction its height of 92 m was the tallest in the world. In order to cope with heavier loading the bridge was replaced by a steel structure in 1900 (see Figure A2.7). The new towers were mounted on the existing masonry pedestals reusing the 1882 anchor bolts. On the eastern face of towers four to 17 the anchor bolts were extended vertically via a collar coupling to accommodate expansion bearings. This was to prove of great importance in later events.

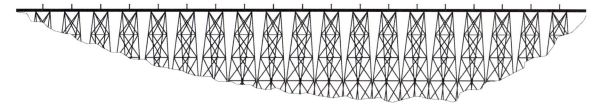

**Figure A2.7**    *Profile of Kinzua Viaduct*

The bridge remained in service until 1959. In 2002 an in-depth inspection found that it was at risk from strong winds and it was closed to all users. From external observation it was assumed that the anchor bolts were subject to reasonable corrosion and did not require replacement.

Maintenance and repair work to the towers and secondary members was started in 2003. On 21 July 2003 a tornado with wind speeds exceeding 90 mph struck the gorge. The windward side of the structure experienced uplift to the extent that the anchorage system was insufficiently strong and 23 of the 41 spans collapsed.

A board of inquiry was set up which concluded that anchor bolts at the base of the collapsed towers had failed rapidly by one of two modes of separation, resulting in segments of the structure becoming airborne under the action of the wind (Gannett Fleming, 2003). Failure occurred in the 1882 anchor bolts which were the weak link in the structure. In designing the strengthening scheme it had been assumed from external examination that strengths of the anchor bolts were within commonly accepted values for wrought iron with reasonable account taken for corrosion.

Two failure modes were identified:

1  **Failure mode 1** in the collar couplings was responsible for 75 per cent of the failures. All the collar couplings seen at the site exhibited radial cracking. The

fractures exhibited evidence of fatigue with final fractures due to overload during the collapse. Many of the couplings had already been fractured by fatigue before the collapse. Inspection beforehand would have been handicapped, if not made impossible, by the presence of washers surrounding the couplings and hiding the cracks from view. The cracks could only have been inspected if the restraining bolts and washers were removed.

2  **Failure mode 2** in the anchor bolts was responsible for 25 per cent of the failures, Figure A2.8. The original 1882 anchor bolts were found to have ductile fractures caused by tensile overload. The bolts were severely corroded and section losses of 20 per cent were assumed for the purposes of structural analysis.

The pre-existing fatigue cracking, within approximately 75 per cent of the collar coupling assemblies, had reduced the uplift capacity of the anchor bolt assemblies to marginally low levels.

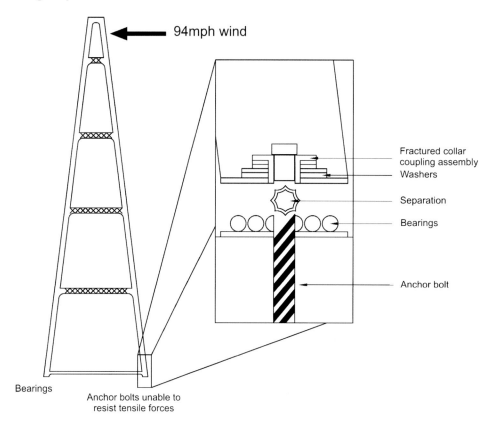

**Figure A2.8**      *Failure mode 2 in anchor bolt of Kinzua Viaduct*

A laboratory study was carried out at Lehigh University by Kaufmann and Connor (2003). The 32 mm diameter bolts from the original 1882 structure were found to have a microstructure and chemical composition typical of the wrought iron of the time. The collar and 38mm bolt added in 1900 were determined to be wrought iron and mild steel respectively. Tensile strengths estimated from hardness measurements ranged from 376 N/mm² to 424 N/mm². Corrosion losses in the original 32 mm bolts varied from being negligible to approaching full section at locations below the foundation surface.

Fractures leading to the collapse had occurred at locations which were hidden from view.

A full analysis of the collapse, including the laboratory investigation at Lehigh University is given in the *Report of the board of inquiry* (Gannett Fleming, 2003).

CIRIA C664

# A2.4      BRAEMORE GREAT BRIDGE, HAMPSHIREE

> Strengthening a steel lattice girder bridge by replacement of corroded components and modification of the structural action.

Braemore Great Bridge is a Grade II listed structure across the environmentally sensitive River Avon in Hampshire. Built c1900 it is a fine example of an early steel lattice girder with transverse trough decking. A 7.5 ton weight limit was imposed in the late 1990s after an inspection revealed extensive corrosion, particularly in the bottom flanges of the girders, shown in Figure A2.9, and pier support beams. A structural assessment also revealed that without U-frame action the girders, even in good condition, could be considered capable of supporting only a 7.5 ton live loading. The deck itself, although corroded, had been strengthened in 1980 with reinforced concrete infill. The corroded deck troughing is shown in Figure A2.10.

**Figure A2.9**      *Corrosion at the bottom flanges*

**Figure A2.10**      *Corrosion of the steel troughing of the main girders*

Work on the bridge was subject to a number of constraints:

- as a listed structure it was necessary to retain as much of the fabric of the existing bridge as possible

- any strengthening had to be sympathetic to the existing structure.

- the heavily reinforced concrete deck made access to the inner faces of the diagonal members impractical

- the fast flowing river, subject to rapid rises after rainfall, made working potentially hazardous

- jacking up and working *in situ* was not considered feasible as the ends of the bridge were built into the abutments.

The area is a site of special scientific interest (SSSI), a special area of conservation (SAC) and a site of importance for nature conservation (SINC).

Prior to starting work on site, environmental investigations were carried out to check for the presence of otters, water voles, pea mussels and Desmoulin's Whorl snails. Botanical and river corridor studies were also carried out.

As it was not possible to work on the bridge *in situ* it was decided to remove it completely and restore it under workshop conditions. The surfacing and structural concrete were broken out to reduce the weight, and the spans were individually lifted into a field where the severely corroded trough decking was cut away. The girders were transported to a fabricator's workshop.

Notable features of the work were:

- the bottom flanges were completely replaced on all four girders. In order to reduce maintenance problems the flanges were made from structural T's welded to plate rather than replacing the original back-to-back angles which had acted as moisture traps and suffered badly from corrosion

- the bottom 300 to 500 mm of most diagonal members had to be removed and replacement pieces connected with full penetration butt welds. Also, about 40 sections were replaced at cross-over points. The original rivets passing through the spacer blocks were replaced with specially made dome headed bolts to retain the original appearance

- all connections were welded and false rivets were welded onto replacement members in the same configuration as on the original connections

- structural analysis of the decorative scrolls was undertaken. By welding the connections and with minimal modifications to the steelwork, it was possible to turn the decorative scrolls into frames strong enough to resist U-frame forces, see Figure A2.11. The new structural elements of these frames were either shaped to be in sympathy with the originals, or kept below road level so as not to be visible

- trough decking was specially pressed to replicate the profile of the original. Thicker sections were used to satisfy current loading criteria and the sections were welded together to minimise corrosion problems at joints

- the girders were placed on bearings at the abutments to enable better access to the undersides for future maintenance

**Figure A2.11**

*Structural U-frame scroll and welded bottom flange connection showing false rivets*

**Figure A2.12**    *The repaired bridge*

- relatively weak lightweight concrete was used on the deck itself with a nominal mesh to enable it to be broken out comparatively easily should the need arise in the future. This was then waterproofed

- the troughing was sprayed with aluminium, and both girders and troughing were coated with a glass flake paint system to provide maximum protection with minimum maintenance. The restored bridge is shown in Figure A2.12.

The whole project lasted approximately 30 weeks, the bulk of the time being spent on fabrication. A footbridge was provided for the duration of the works for pedestrians and cyclists.

# BROCKHAMPTON BRIDGE, HAMPSHIRE

> Heat straightening repair of a bridge having steel plate girders damaged by vehicular impact.

Brockhampton Road New Bridge is a three-span structure carrying a local road across the A27 trunk road in Hampshire. The deck is composed of six Grade 50 steel plate girders supporting a 210 mm thick reinforced concrete slab. Each main beam is continuously braced in pairs and is 1000 mm deep with a 480 mm wide bottom flange and a 300 mm top flange. The spans lengths are 14.9 m, 34.6 m, 14.9 m.

In July 2002 a low-loader lorry carrying heavy machinery eastbound on the A27 struck the underside of the bridge causing structural damage to the main beams and bracing (see Figure A2.13). At this position the bottom flange thickness was 25 mm and the web thickness was 12 mm at the point of impact. The damage occurred at the point of contraflexure in an area where the bottom flange was in tension. It was not considered necessary to impose traffic restrictions.

**Figure A2.13**     *Impact damage to flange of main beam*

The trailer had been loaded to the correct height but impact occurred due to inadequately secured machinery bouncing as the vehicle moved. All the impact damage occurred before the driver had time to bring the vehicle and trailer to a stop.

Repair work was constrained by the need to close the east bound carriageway of the A27 which is heavily trafficked by up to 82 000 vehicles per day. Diversions were acceptable only between 2100 and 0500 hours.

After consideration of different methods of repair, it was decided to carry out repair work as follows:

- repair the distorted bottom flanges of beams one and five by heat straightening. It was estimated that the alternative of replacing the beams would be five times the cost

- reinstate the buckled stiffeners, K-bracing and cracked welds

- apply protective coatings of paint on all affected areas.

The heat straightening was carried out in accordance with guidance provided by the Federal Highway Administration (US DoT, 1998). In heat straightening, concentrated heating is applied locally to the distorted area causing the steel to expand. This expansion is restrained by the cooler surrounding metal causing the heated area to yield in compression. On cooling, the affected area contracts so that by using carefully designed heating patterns, plus external jacking, the beam can be returned to its original profile.

**Figure A2.14**     *Repair work on main beam*

The affected areas of steel were inspected using magnetic particle testing during stages of the work and on completion. No cracking or other defects were detected.

The repairs were completed without undue disruption to the traffic or local residents.

For more detailed information about the structural analysis and design of the repair, see Clubley *et al* (2006).

# BATTERSEA YARD RAILWAY BRIDGE, SOUTH LONDON

> Strengthening a wrought iron rail bridge by infilling beneath the deck and converting to an embankment.

Battersea Yard Railway Bridge, built in 1866, has five spans of wrought iron girders supported on heavily skewed brick piers. The bridge spans 120 m over Battersea Yard Depot and carries three rail tracks on sleepers fixed to longitudinal baulk timbers. It was identified as requiring repair and strengthening work to provide a 30 year extension to its life. Various options were studied ranging from complete renewal to repair and strengthening the components in need of attention. The solution finally adopted was to infill and convert the bridge to an embankment. This enabled work to be undertaken without affecting the train service or line speed. A detailed account of this work is provided by Webster and Mehrkar-Asl (2002).

**Figure A2.15**      *Infilling work near to completion*

The design comprised a lightweight pulverised fuel ash (PFA) and foam concrete fill embankment supported on a grid of over 1000 bored and continuous flight augered piles of 300 to 400 mm diameter, and between five and 19 m long. The advantages of this scheme were:

- significantly lower cost
- minimum requirement for track possessions
- lower future maintenance costs.

The main requirements of the design were to limit settlements to acceptable values and ensure that they did not affect an adjacent rail viaduct a few metres away.

Driven piling was ruled out due to the risk of vibrations causing damage to the adjacent viaduct and the potential for long-term settlement. Bored cast *in situ* pile construction was used beneath the arches due to restricted headroom and a continuous flight auger rig was used in other areas. The friction piles were located at either end of the viaduct to limit differential settlements, and the base capacity piles were located away from the abutments. In order to meet the requirements, two pile types were used:

1    A longer pile which shed load via shaft friction through the London clay

2    A shorter pile which shed load via the base and into the gravels.

The load distribution layer was 800 mm thick crushed concrete reinforced with an unsheathed structural geotextile. The infill was composed of PFA laid onto the granular distribution layer and up to the undersides of the bridge girders. PFA was adopted to minimise the embankment loading and because of its high strength gain following deposition. The PFA slopes were reinforced by Geogrid Tensar. Durability was provided by topsoil and grass on the east elevation, and C30 sprayed concrete containing polypropylene fibres to combat water runoff from the adjacent structure on the west side. The rest of the embankment was filled with foamed concrete designed to transfer load from the ballasted track to the PFA. A minimum 28-day strength of 2 $N/mm^2$ and density of 1400 $kg/m^3$ was specified to satisfy strength requirements while restricting embankment loading. Foam concrete provided a reasonably fast method of filling the inaccessible areas between the girders.

Structure monitoring was carried out during the construction period, when 60 per cent of the predicted settlement was expected to occur. Measurements were taken at regular intervals:

- targets were positioned on the main girders and survey pins were fixed to the masonry piers, which were read from a base station located outside the critical settlement zone.

- verticality of the piers was measured to detect any possible rotation.

- inspection and measurement of existing cracks in the piers and abutments

- track levels were taken to check vertical alignment, cant and twist

- regular inspections of the structure were made during the works

- visual track inspections

- track geometry was checked by a high speed track recording machine.

Work started in 1998 and was completed in 2001. The completed embankment is shown in Figure A2.15. For more detailed information about this work see Webster and Mehrkar-Asl (2002).

Corrosion performance and maintenance painting of a historic cast iron canal aqueduct:

• during its 200-year life the structure was previously painted on only three occasions; in 1886, 1936 and 1965

• in 2000, the ironwork was found to be in excellent condition with little loss of material due to corrosion

• in carrying out the maintenance work, conservation principles were followed and repairs and painting were carried out using traditional materials and methods even where there was added expense.

Pontcysyllte Aqueduct was designed by Thomas Telford and built between 1795 and 1805 to carry the Ellesmere Canal across the valley of the River Dee. A view of the structure is shown in Figure A2.16. It is widely regarded as the most spectacular achievement of waterway engineering in Britain and a pioneer of cast iron construction. It is designated as a scheduled ancient monument and a Grade I listed structure. It is on the tentative list for World Heritage Sites. The River Dee has international significance and is designated as a site of special scientific interest. It is also a candidate special area of conservation.

The aqueduct spans 307 m on 18 masonry piers at a height of 38.4 m above the River Dee with four of them in the river itself. The canal is in an iron trough supported by 19 arches of 13.8 m span. The troughs and arches are assembled from cast iron sections bolted together with wrought iron fittings. The trough is 3.6 m wide and 1.7 m high. The towpath extends over the water surface for 1.2 m and there is an iron handrail and balustrade to protect the towpath with identical sections on the opposite side over the abutments. The offside of the trough has no other rail protection.

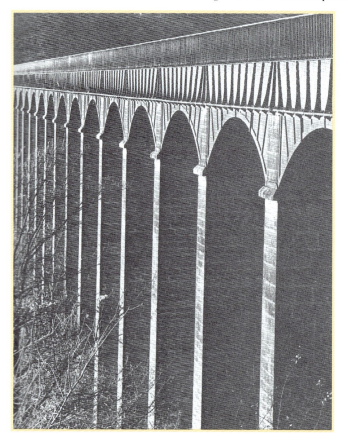

**Figure A2.16**

*Pontcysyllte Aqueduct*

The first reliable evidence of painting the ironwork was not until 1886, ie 81 years after construction. It was repainted in 1936 and in 1965 when a bitumastic paint was applied to the outside elevations of the troughs and arches. Coal tar is the only painting material mentioned in documentary research before 1965. Despite these long intervals between painting maintenance, apart from the bolts there has been very little loss in material due to corrosion.

In 2000, trials were carried out on three protective systems prior to refurbishment for the 200th anniversary of the structure:

1   Grit-blast preparation, two coats of Zinga galvanic protector, two coats of polyurethane.

2   Grit-blast preparation, one coat of coal-tar epoxy.

3   Wire-brush preparation, one coat of red lead primer, one of coat bitumen.

The report on the trials concluded that all three systems had limitations, but grit-blasting and painting would give the longest life to first maintenance.

From a conservation perspective there are particular problems in grit-lasting historic cast iron as it is not a homogeneous material:

● most of the ironwork was cast in open floor mouldings which impart a hard surface to the exposed faces in addition to the overall protective fire skin. These surfaces also contain defects and are likely to be adversely affected by grit blasting

● blasting is likely to remove historic material used to fill casting flaws that would have corroded into a solid mass

● blasting would not remove heavy rust scabbing which would require inspection and needle gunning or hand chipping

● blasting would drive out the caulking of dried hemp from flange joints and expose the white lead sealing material. Replacement with a modern sealant would have an expected life of about 15 years.

It was concluded on heritage grounds that the structure should not be grit-blasted. The ironwork had survived for nearly 200 years and it was difficult to justify a process that would cause irreversible damage. The cost of wire brushing would be an acceptable price to pay for the integrity of such a significant monument.

The painting scheme selected was:

● primer coat on areas where the original paint had completely failed, exposing partially corroded bare metal. This accounted for approximately 50 per cent of the surface area

● three coats of bitumen paint with a total DFT of 300 microns

● coatings were brush applied to internal surfaces due to the congestion of follow-on trades which prevented areas from being effectively isolated

● coatings were sprayed onto 60 per cent of external areas where it was possible to segregate the operation from other trades.

Some of the iron bolts securing sections together within the trough had suffered considerable corrosion. In the trough the damage was largely to the nuts and heads rather than necking of the shanks. The captive bolts on the arch panels were more seriously corroded. Heritage issues dictated that it was preferable to re-forge recovered material allowing thread pitch and the square head design to be retained. New bolts were date stamped to record their introduction. Modern bolts such as galvanized bolts with round heads and hexagonal nuts were avoided as they would not be in sympathy with the original construction. Finally, if there was a problem replicating the original lead based washers alternative traditional sealers were used in preference to modern materials such as nylon or PTFE washers.

## A2.8     ARMSTRONG BRIDGE, NEWCASTLE

> Repair of a historic wrought iron bridge having severe corrosion.

Armstrong Bridge is a Grade II listed structure across the steep valley of the Ouseburn at Benton Bank, Newcastle, see Figure A2.17. It was named after Sir William Armstrong who provided most of the construction costs and was opened to traffic in 1878. It is constructed in wrought iron and has eight spans of equal lengths, a total length of 552 ft (168 m) and a usable width of 25 ft (7.6 m). The superstructure was originally fixed at the abutments and had an expansion joint at the centre. The spans were supported on cast iron rocker bearings to enable thermal movements and settlements due to mining subsidence to be accommodated, shown in Figure A2.18. The piers were formed of pairs of columns, also supported on rocker bearings, and connected laterally by pairs of cross-bracing tie rods. Longitudinal lattice girders were supported on box girders resting on the piers. The deck was formed using domed iron plating.

In 1906, only 28 years after construction, extensive corrosion was found to have occurred in four of the 14 columns where the wrought iron was covered by ornamental castings and water had collected at the interfaces. In one column the wrought iron was corroded to the extent that daylight could be seen. The castings were removed, maximum permissible traffic loads were reduced from 12 to 6 ton, and a 5 mph speed limit was imposed. However, there was no watchman and it was believed that these limits were flouted.

**Figure A2.17**     *Armstrong Bridge*

**Figure A2.18**     *Iron rocker bearings*

The bridge performed well and there was no significant maintenance reported in the next 50 years.

- in 1956 maximum permissible loads were reduced from six to two tons
- in 1960 urgent repairs were made when it was found that timbers supporting the footpath had rotted
- in 1963 the bridge was closed to vehicular traffic and limited to pedestrians.

In the 1970s the bridge was surveyed and various problems became apparent:

- there was corrosion in secondary girders beneath the roadway and in the columns
- some of the cast iron pilasters were cracked
- the central expansion joint was corroded and locked solid
- failure of the expansion joint had transferred movement to the abutments causing damage so that the western abutment required shoring
- wet rot was present in many of the timbers.

Costs for addressing these problems were estimated as follows:

- major repairs to the bridge, including painting: £195 000
- construction of a new concrete footbridge (narrower than the original bridge): £25 000
- construction of a replica bridge: £600 000
- demolish the bridge: Not costed.

Comments were invited from the general public, and here was overwhelming support for retaining the bridge as it was seen as serving a useful purpose as well as having historic associations and being an attractive structure. It has now become popular as an open-air weekend art market.

A refurbishment scheme was prepared which included changing the articulation by clamping the central expansion joint, tying back the east abutment to reinforced concrete anchor walls and installing sliding bearings in the west abutment. The wrought iron columns were replaced with tapered I-section steel columns and the original wind bracing members were shortened and re-fitted, see Figure A2.19. Decorative cast iron panels and top flange pilasters masking the joints between longitudinal beams above the columns had weathered badly, and few were still in position. A new set made in fibreglass was commissioned and fitted at a later stage in the planned refurbishment. Some of the background and the first stages of the work have been published, see Restoration (1983) and Bussell (1984).

**Figure A2.19**

*Wind bracing*

The last stages of the work were completed in 1993–94 when the deck was waterproofed and resurfaced. Drainage was improved, lighting was installed and the structure was repainted.

The bridge receives a general inspection every two years and a principal inspection every six years. Although some slight water seepage was detected, no significant deterioration had occurred in the subsequent 12 years. The bridge is frequently used, and has attracted little graffiti and no serious vandalism.

The lessons learned from this case study are:

- ornamental castings placed over the columns might have appeared decorative but allowed corrosion to occur in places not visible or readily accessible for inspection

- ingress of debris and corrosion occurred in the central expansion joint which was below the deck and not readily accessible for maintenance, and locked up.

## A2.9      BALGAY PARK FOOTBRIDGE, DUNDEE

> An example of unsuccessful repair and strengthening work on an iron footbridge.

Balgay Park Footbridge, Dundee, shown in Figure A2.20, is an elegant structure built in 1872–73. It was originally constructed in cast and wrought iron. It has a central arch of 25.6 m span and two simply supported side spans of 5.9 m. The central span crosses a pedestrian footpath at a height of about 13 m and comprises two arch ribs supporting longitudinal edge beams. The ribs are segmental, each composed of three castings bolted together at approximately the third points. They are connected transversely by cylindrical members with tapered ends bolted to the ribs via a flange. The two side spans are between the outer edges of the masonry piers and bankseats at the ends of the structure. Edges of the deck are protected by decorative cast parapet panels secured to the top flange of the edge beams with bolted connections. It is surmised that the bridge originally had a wooden deck.

**Figure A2.20**      *Balgay Park Footbridge*

**Figure A2.21**     *Orthogonal disposition of RSJs and location of transverse connections*

In the early 1920s, and again in 1929, maintenance and strengthening work had been carried out:

* the wooden deck was replaced by a 75 mm thick reinforced concrete slab cast onto the steel joists. The slab contains a single layer of mild steel mesh having a diamond pattern and composed of rectangular section bar

* an assemblage of transverse rolled steel joists was inserted and attached to the longitudinal edge beams, or arch ribs, depending on position. These support a row of longitudinal joists, bolted to the webs of the transverse joists centrally on the longitudinal axis of the bridge (Figure A2.21)

* the transverse connections between the joists and edge beams were buried in concrete encasing the internal sides of the cast iron edge beams.

In the time since the repairs were carried out the steel and concrete had weathered very badly, whereas the original ironwork remained in good condition.

The steel joists are in very poor condition. Although they appear to have been painted in the past, the paint system has largely failed and the steel is badly corroded. The corrosion is particularly severe at the ends of the transverse joists and along the top flanges of both longitudinal and transverse joists. In addition to the holes in the webs, the top flanges have disintegrated at several locations with evidence of significant expansive corrosion as shown in Figure A2.22.

**Figure A2.22**   *Corrosion of transverse steel*

**Figure A2.23**   *Exposed and corroded joist reinforcement*

The deck slab is in poor condition having several through-thickness cracks across its full width. The cover to the mesh reinforcement is minimal in places and there are several areas on the soffit where the mesh is exposed (Figure A2.23). At the majority of these areas the mesh has completely corroded away.

The concrete encasing the edge beams is in very poor condition. There is systematic cracking present along the interface between the top of the concrete encasing the inside of the edge beams and the deck. This cracking, which is wider than 10 mm in places, has resulted in the complete spalling of the encasing concrete at some locations. Where the ironwork is exposed it is in reasonable condition.

Overall the repairs are deemed to have been unsuccessful as the new material weathered badly and was less durable that the original ironwork. This was due to poor design and execution of the work.

There are lessons from this case study:

- when concrete and steel are added to an iron structure, regular maintenance becomes more important
- water management and waterproofing are very important
- interfaces between concrete and steel are especially susceptible to corrosion and should be avoided where possible. Unavoidable interfaces should be thoroughly waterproofed and be accessible for inspection and maintenance.

The bridge is scheduled to be renovated and returned to its former state.

## A2.10　KILLEARN STATION BRIDGE, STIRLINGSHIRE

> Reconstruction of a bridge with limited load carrying capacity and weakened by corrosion, to enable it to meet full loading requirements.

Killearn Station Bridge is located to the west of the village of Killearn in Stirlingshire and owned by BRB (Residuary) Ltd. It carries a minor road, the B834, over a former Strathendrick and Aberfoyle Railway, built around the 1880s. After the line was closed in 1959, this section was converted into a footpath for the West Highland Way in 1980. Following assessments, a temporary weight restriction of three tons was imposed on the bridge in 2004 by Stirling County Council. The assessment showed that the main girders failed under both shear and bending due to severe corrosion, and the buckle plates had a limited live load capacity.

The original structure comprised seven riveted wrought iron girders with buckle plates, curved in both directions. The substructure is of sandstone and the parapets were wrought iron panels fixed to the top flange of the external girders, and to the stone pilasters which sat on top of the external girders.

Strengthening the bridge was subject to various constraints:

- support was required for a 6 inch (152 mm) steel water main and underground telecommunication cables including fibre optics
- a site of special scientific interest (SSSI) was located immediately to the south west of the structure
- it was required that the existing headroom of 2.3 m should be retained.

A feasibility study was undertaken to bring the structure up to current loading standards and the following options were proposed:

**Option 1**　Infilling and providing an underpass. The existing water main would be bridged by precast concrete planks supported on brick walls. The void between the underpass and the bridge abutments would be filled with foamed concrete and the exposed concrete would then be faced with masonry. Minor masonry repairs to the pilasters would be carried out and the corroded parapets would be treated and repainted.

**Option 2**　Propping the existing girders, making remedial repairs to the bridge structure, and installing a concrete slab over the buckle plates. With the additional restraint in place and the girders acting as two-span continuous elements, the superstructure would have the required load carrying capacity.

**Option 3**　Reconstruction, comprising the removal of the existing superstructure and replacing it with a new deck of similar dimensions. The substructure was to be repaired as necessary.

**Option 4**　Infilling, removing the superstructure and lowering the road level by 2 m, incorporating ramps for pedestrian and cycle access over the B834. The water main would be bridged as in Option 1.

The infilling Options 1 and 4 were rejected due to high cost and risks associated with protecting the water main, and restrictions to the footpath users.

The most economic solution was provided by Option 2 but the repairs could only guarantee 25 years without further major intervention.

Although not the most economic solution, Option 3 reconstruction was considered to be best as it removed the high risks associated with protecting the existing water main under the bridge.

Figures A2.24 and A2.25 show views of the bridge under demolition and reconstruction.

**Figure A2.24**     *Partial demolition showing exposed buckle plates and riveted beam*

**Figure A2.25**     *Partial reconstruction showing temporary support of the services and new steel beam in place*

## The reconstruction scheme

There were no signs of distress in the substructure confirming that no additional strengthening measures were required. The foundations were deep and the backfill material was of good standard.

However the following repairs were required:

- cracks in the abutments were to be stitched using Cintec anchors
- large areas of the mortar joints in the abutments and wing walls were to be raked out and re-pointed
- dislodged coping stones and loose masonry on the wing walls were to be repaired
- vegetation around the wing walls were to be removed and treated.

Taking into account the various requirements, it was decided that a steel-concrete composite deck provided the best engineering solution, and value for money. The existing carriageway width of 5.5 m was to remain the same and 1 m wide surfaced footways would be provided at either side of the carriageway. The new deck was designed as a semi-integral structure to minimise any deterioration of the beams and bridge bearings associated with the provision of expansion joints. The thermal effects were minimal and predicted settlements were very small.

**Figure A2.26**     *The new superstructure*

The work was completed in 2005 and is shown in Figure A2.26.

# A2.11    HOLLOWAY ROAD BRIDGE, LONDON

> A temporary repair designed as a holding action to last until a new deck is constructed.

Holloway Road Bridge, built in 1868, carries the A1 over the Gospel Oak to Barking railway line. It is simply supported on an 8.5º skew and has a square span of 7.3 m. The carriageway is 12 m wide with two lanes for general traffic and a dedicated bus lane. Footways on either side are 3.0 and 3.2 m wide. It was originally constructed with longitudinal cast iron beams with overall depths varying from 355 mm at the abutments to 455 mm at the centre and at 1.2 m centres. The edge parapet beams were also cast iron and 610 mm deep. Curved buckle (hogging) plates supporting clay infill between the beams. The buckle plates were positioned on the lower flanges and bolted to the webs of the beams as shown in Figure A2.27.

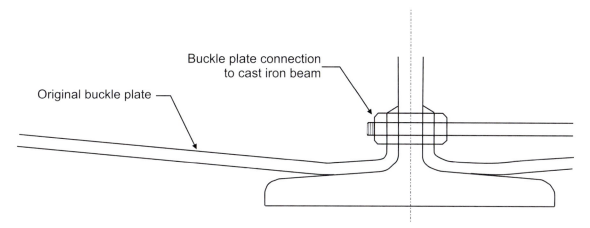

**Figure A2.27**        *Buckle plate connection to cast iron beam*

In 1908, the tramways in Holloway Road were adapted for electric traction and it was necessary to remove 10 of the cast iron girders under the carriageway, and to substitute 13 steel girders at 760 to 1065 mm centres to suit the geometrical requirements of the trams.

In 2000, an inspection revealed that the bottom flange plates of the steel beams supporting transverse RSJs, steel plates and York Stone slabs were badly corroded and many of the heads of the rivets connecting the flange plates to the beams were missing where they had been exposed to exhaust gases from decades of passing steam engines. In addition, one of the York Stone slabs was cracked. The buckle plates spanning between the remaining cast iron footway beams were in a particularly poor condition with some sections completely corroded and exposing the clay fill material above.

The superstructure was assessed to BD 21/01 taking into account the findings of the inspection. The main steel beams were found to be adequate for 40 ton loading. However the cast iron beams under the footways were assessed as not being capable of supporting accidental wheel loading, and the risk of fill material falling onto the tracks due to loss of support to the transverse elements and corrosion of the hogging plates between the cast iron footway beams was considered a hazard to the safety of the railway. Immediate measures were implemented by overslabbing the carriageway area with a 150mm thick concrete slab bearing on the beams, or fill directly above the beams but separated from the fill between the beams with a compressible filler.

Shallow services (eg electricity, gas, water supply etc) in the verges prevented the slab being extended under the footways, so it was not possible to provide similar protection to accidental wheel loading on the fill between the cast iron beams. Furthermore, although it was considered likely that there would be some reserve of capacity, the cast iron beams, when assessed in accordance with BD 21/01, would not be capable of supporting the applied loading.

It had been decided to replace the entire deck. A design has been prepared and is programmed to be constructed in about 10 year's time. In the mean time, parking on the verges has been prevented by installation of open box beams. This measure was not considered to be effective in the event of an accident. More substantial repairs have been carried out to ensure that the bridge can be used with safety until the new deck has been constructed.

Due to the presence of the services, it was not possible to carry out any of the work from above the deck and in consequence everything was done from below, see Figure A2.28.

**Figure A2.28**     *Steel plates positioned from below the bridge*

The temporary repair involved the following works:

- steel plates were manoeuvred into place beneath the buckle plates and above the bottom flanges of the cast iron beams. The plates were in short lengths to enable this handling, and were keyed together longitudinally

- for some of the beam spacings it was necessary to cut the ends of the buckle plates and seat the new steel plates on resinous bedding mortar as shown in Figure A2.29

- mortar was packed into the voids between the buckle plates and new steel plates.

Preliminary calculations indicated that the thickness of plates required to support full accidental wheel loading would be so heavy as to theoretically overload the cast iron beams. The plates were therefore not designed for full accidental wheel loading but for a combination of loads with effects that would be commensurate with the capacity of the cast iron beams including self-weight, the dead load of the mortar and the fill currently supported by the buckle plates, in combination with pedestrian live loading of 5 kN/m². The final thickness of the plates was 10 mm.

The works were carried out during a 52 hour possession of the railway and were planned to ensure that the tracks were cleared two hours before the scheduled re-opening of the line. In the event, work progressed very well and was completed 17 hours early. Live traffic over the structure was maintained at all times.

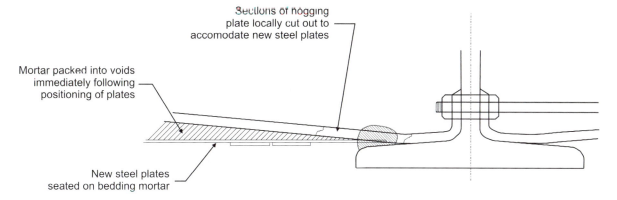

**Figure A2.29** *Location of steel plate on bottom flange*

The completed repair work is shown in Figure A2.30.

**Figure A2.30** *Completed repair work*

# A3  Industrial coating applicator training scheme (ICATS)

The Institute of Corrosion set up a training and qualification scheme for paint operatives. This initiative was launched in 2005, and has been funded and supported by the Highways Agency, Network Rail and the CITB. The programme has also been verified by CITB.

This qualification has three stages where operatives attend full time training over a period of three weeks. To achieve the top level, Level III, at least two years practical inspection experience must be undertaken before taking written, oral and practical examinations set to a high standard. The Institute of Corrosion maintains a register of successful candidates and renewal is required at five-yearly intervals. It is important for any bridge owner or major structure consultant to ensure that the right level of qualification for the inspection requirement is put in place during the pre-contract negotiations.

Among the items covered in the basic course are:

- health and safety
- site access
- plant and equipment
- surface preparation
- paint types and application
- quality control.

Upon completion of this course, there will be the opportunity to attend any of four specialist options, namely:

- abrasive blast cleaning operator
- paint sprayer
- thermal (metal) sprayer
- water jetting operator.

The scheme is known as the Industrial Coating Applicator Training Scheme (ICATS). Contractors register with the Institute of Corrosion and nominate key members of their staff to become qualified coating trainers. They then register each individual on their staff who will undertake 39 hours of formal training in 64 modules (theory and practical). Each painter who is successful in passing this qualification will be entered onto the Institute of Corrosion's ICAT register.

At the time of writing, this ICATS (Industrial Coating Applicator Training Scheme) will be phased in by the Highways Agency Skills Sector route during 2007 and 80 per cent of operatives will require registration by January 2008. The Environment Agency are also introducing the scheme for painting contractors during 2007.

# A4　　　Bibliography

## A4.1　　MATERIALS AND STRUCTURAL FORMS

ANGUS, H T (1976)
*Cast iron: Physical and engineering properties (2nd edition)*
Butterworth & Co Ltd, British Cast Iron Research Association, London, (revised 1978),
(ISBN 0-40870-933-2)

BCF LIMITED (1998)
*Old lead painted surfaces: A guide on repainting and removal for D-I-Y and professional painters
and decorators*
British Coatings Federation Ltd, Surrey. Available from: <http://www.coatings.org.uk/>

BADOUX, M and SPARKS, P (1998)
"Fracture critical study of an historic wrought iron bridge"
*Structural Engineering International*, no 2, pp 136-139

BATES, W (1984)
*Historical structural steelwork handbook (Covers cast and wrought iron)*
BCSA, London (ISBN 0-85073-015-5). Available from: <http://www.steelconstruction.org/>

BRÜHWILER, E, SMITH, I F C and HIRT, M A (1990)
"Fatigue and fracture of riveted bridge members"
*Journal of Structural Engineering*, vol 116, **1**, January, pp 198–214

CARR, J C and TAPLIN, W (1962)
*History of the British steel industry*
Harvard University Press, USA

CULLIMORE, M S G (1967)
"The fatigue strength of wrought iron after weathering in service"
*The Structural Engineer*, vol 45, **5**, May, pp 193–199

EVANS, U R (1960)
*The corrosion and oxidation of metals: scientific and practical applications*
Hodder Arnold Ltd, London (ISBN 978-0-71312-054-7)

GALE, W K V (1964)
"Wrought iron: a valediction"
*The Newcomen Society*, vol 36, pp 1–11

GALE, W K V (1965)
"The rolling of iron"
*The Newcomen Society*, vol 37, pp 35–46

GALE, W K V (1967)
*The British iron and steel industry: a technical history*
David & Charles, Newton Abbot, UK (ISBN 0-71534-988-0)

GALE, W K V (1969)
*Iron and steel*
Longmans, London

GALE, W K V (1971)
*The iron and steel industry: a dictionary of terms*
David & Charles, Newton Abbot, UK

GALE, W K V (1974)
"The Bessemer steelmaking process"
*The Newcomen Society*, vol 46, pp 17–26

GALE, W K V (1981)
*Ironworking, Shire Album 64*
Reprinted 1994, Shire Publications Ltd, Oxford, UK

HALLIWELL, S M (2000)
*Polymer composites in construction*
BRE Report 405, Building Research Establishment, Garston

INSTITUTION OF STRUCTURAL ENGINEERS (1999)
*Guide to the structural use of adhesives*
Institution of Structural Engineers <http://www.istructe.org/index.asp?bhcp=1>

JAMES, J G (1981)
"The evolution of iron bridge trusses to 1850"
Reprinted 1997, *The Newcomen Society*, vol 52, pp 67–101 Sutherland, R J M (ed)
*Structural iron, 1750-1850: Studies in the history of civil engineering*, vol 9, pp 311–345,
Ashgate Variorum. Available from: <http://www.newcomen.com/>

JAMES, J G (1988)
"Some steps in the evolution of early iron arched bridge designs"
*The Newcomen Society*, vol 59, pp 153–186

KEATING, P B *et al* (1984)
"Fatigue behaviour of welded wrought iron bridge hangers"
*Transportation Research Record*, vol 2, no 950, pp 113–120

MITCHELL, D S (2001)
*General guidance notes for cast iron*
Heritage Engineering Limited, Carstairs

MOY, S S J (ed) (2001)
*ICE design and practice guide: FRP composites – life extension and strengthening of metallic structures*
GAR011, Thomas Telford Publishing, London (ISBN 978-0-72773-009-1)

REES, G I (1999)
*Welding of cast irons – a guide to best practice*
The Welding Institute, Cambridge

ROYAL COMMISSION (1850)
*Report of the Commissioners appointed to inquire into the application of iron to railway structures*
Her Majesty's Stationary Office, London

RUDDOCK, E (1979)
*Arch bridges and their builders: 1735-1835*
Cambridge University Press (see Chapters 11-13 for iron arch bridges)
(ISBN 0-52121-816-0)

RUDDOCK, E (2003)
"Some iron suspension bridges in Scotland 1816-1834 and their origins"
*The Structural Engineer*, 4 March, pp 23–29

SHANMUGANATHAN, S (2003)
"Fibre reinforced polymer composite materials for civil and building structures – review of the state of the art"
*The Structural Engineer*, 1 July, pp 26–33

SMITH, D (1992)
"The works of William Tierney Clark (1783-1852), civil engineer of Hammersmith" (includes discussion of concerns relating to potential boat race crowd loading in 1869 on Hammersmith suspension bridge)
*The Newcomen Society*, vol 63, pp 181–207

SUTHERLAND, J (1992)
"Cast iron"
*Construction materials reference book*, Doran, D K (ed) pp 3/1–3/15, Butterworth-Heinemann, London

SUTHERLAND, J (1992)
"Wrought iron"
*Construction materials reference book*, Doran, D K (ed), pp 4/1–4/12, Butterworth-Heinemann, London

SWAILES, T (1996)
"19th century cast iron beams: their design, manufacture and reliability"
*Civil Engineering*, Institution of Civil Engineers, London, vol 114, **1**, February, pp 25–35

SWAILES, T and DE RETANA, E A F (2004)
"The strength of cast iron columns and the research work of Eaton Hodgkinson (1789-1861)"
*The Structural Engineer*, 20 January, pp 18–23

TOPP, C (2001)
*Dissertation on wrought iron (puddled iron)*
Wrought Iron Advisory Centre <http://www.realwroughtiron.com/wiac.asp>

TWELVETREES, W N (1900)
*Structural iron and steel*
Fourdrinier

TYLECOTE, R F (1992)
*A history of metallurgy*
Maney Publishing, London (ISBN 978-1-90265-379-2)

## Standards

BS 5493:1977 *Code of practice for the protective coating of iron and steel structures against corrosion* (in process of being superseded by various parts of BS EN ISO 12944, Paints and varnishes: corrosion protection of steel structures by protective paint systems)

## A4.2 ASSESSMENT

BUSSELL, M (1997)
*Appraisal of existing iron and steel structures*
SCI publication 138, Steel Construction Institute, Berks, UK (ISBN 978-1-85942-009-6)

BUSSELL, M (1984)
"Armstrong Bridge"
*Trust,* Tyne and Wear Industrial Monuments Trust, Nov 198, pp 13–17

CLARK, K (2001)
*Informed conservation*
English Heritage, Swindon (ISBN 978-1-87359-264-9)

H M RAILWAY INSPECTORATE (1996)
*An assessment by HSE (Health & Safety Executive) of the structural integrity of the Forth Rail Bridge*
Her Majesty's Stationary Office, London

HISTORIC SCOTLAND (2000)
*A guide to the preparation of conservation plans*
Historic Scotland, Edinburgh

INSTITUTION OF STRUCTURAL ENGINEERS (2005)
*Appraisal of existing structures*
2nd edition, Institution of Structural Engineers <http://www.istructe.org/>

JONES, D S and OLIVER, C W (1978–1980)
"The practical aspects of load testing"
*The Structural Engineer*, vol 56A, **12**, December 1978, pp 353–356 and discussion vol 58A, **8**, August 1980, pp 251–253

LEEMING, W F (1974)
"Stress investigation and strengthening of the Royal Albert Bridge, Saltash"
*Civil Engineering,* Institution of Civil Engineers, London, vol 56, **1**, November, pp 479–495 and discussion (1976) vol 60, **1**, February, pp 163–174

MITCHELL, G R (1954)
*National Building Studies Research Paper No. 19: dynamic stresses in cast iron girder bridges*
Her Majesty's Stationary Office, London

MORGAN, S K and HEATHORN, T J (1981)
"A study of the design and construction and a structural analysis of Magdalene Bridge, Cambridge: 1 Design and construction (Morgan); 2 A structural analysis (Heathorn)"
*The Structural Engineer*, Vol 59A, August, pp 255–262

TILLY, G P (2002)
*Conservation of bridges: a guide to good practice*
Spon Press, London (ISBN 978-0-41925-910-7)

WARDLE, J B and LUCAS, J C (1975)
"Britannia Bridge: stress investigation before and after the fire"
*Civil Engineering,* Institution of Civil Engineers, London, vol 58, **1**, May, pp 175–193

XIE, M, BESSANT, G T, CHAPMAN, J C and HOBBS, R E (2001)
"Fatigue of riveted bridge girders"
*The Structural Engineer*, vol 79, **9**, 1 May, pp 27–36

## A4.3      EXAMPLES OF REPAIR AND STRENGTHENING

AMERICAN SOCIETY OF CIVIL ENGINEERS (1979)
*Repair and strengthening of old steel truss bridges*
ASCE, USA

ANON (1987)
"The Billingham branch bridge"
*Metal construction*, January, pp 19–21 (*Strengthening of first British all-welded steel bridge of 1934*)

BLAKELOCK, R, MUNSON, S R and YEOELL, D (1998)
"The refurbishment of Westminster Bridge: assessment, design and precontract planning"
*The Structural Engineer*, vol 76, **10**, 19 May, pp 189–194

BROAD, R J (1989)
"Restoration of Stanley Ferry aqueduct"
*Conservation of engineering structures*, Thomas Telford, pp 87–94

CROSSIN, J, MARSHALL, G and YEOELL, D (1998)
"The refurbishment of Westminster Bridge: bridge strengthening"
*The Structural Engineer*, vol 76, **10**, 19 May, pp 195–201

DADSON, J (1982)
"Cambridge builds a new bridge within a bridge (Strengthening of 1823 cast iron Magdalene Bridge)"
*New Civil Engineer*, 8 April, pp 28–29

DODDS, N (2003)
"Strengthening a bridge using carbon fibre reinforced plates (Strengthening of a welded steel plate girder railway bridge)"
*The Structural Engineer*, 4 March, pp 17–19

DODDS, N M S, LOCKE, L A and WELSFORD, R N (1995)
"Recent strengthening work to Cleveland Bridge, Bath"
*The Structural Engineer*, vol 73, **5**, 7 March, pp 69–75

HAYWARD, D (1990)
"Centre parting"
*New Civil Engineer*, 23 August, pp 19–21

HOLLOWAY, B G R and WADSWORTH, H J (1977)
"The strengthening of Hammersmith Bridge"
*Civil Engineering*, Institution of Civil Engineers, London, vol 62, **1**, November, pp 585–604 and discussion (1978) vol 64, **1**, August, pp 475–481

HUDSON, A J (1985)
"Strengthening of Magdalene Bridge, Cambridge"
*Highways and Transportation*, vol 7, **32**, July, pp 17–20

HUME, I (1980)
"The Iron Bridge, Shropshire: repainting and repairs 1980"
*Trans*, ASCHB, vol 5, pp 20–23

HUSBAND, H and HUSBAND, R W (1975)
"Reconstruction of the Britannia Bridge – Part I: Design; Part II: Construction"
*Civil Engineering*, Institution of Civil Engineers, London, vol 58, **1**, February, pp 25–26 and discussion, November, pp 639–655

KAY, J R and STIFF, S J (1995)
"The restoration of Crown Point Bridge, Leeds: Part 1. Design and development [Kay];
Part 2. The contract (Stiff)"
*Construction Repair,* September/October, pp 36–41

KUMAR, A (2003)
"Strengthening of London's Lambeth Bridge"
*Structures and Buildings,* Institution of Civil Engineers, London, vol 156, May, pp 151–164

LARK, R J, MAWSON, B R and SMITH, A K (1999)
"The refurbishment of Newport Transporter Bridge"
*The Structural Engineer,* vol 77, **16**, 17 August, pp 15–21

LOWSON, W W (1967)
"The reconstruction of the Craigellachie Bridge"
*The Structural Engineer,* vol 45, **1**, January, pp 23–28 and discussion Vol 45, **8**, August,
pp 287–289

MATTHEWS, A and PATERSON, I A (1993)
"LMR to LRT: restoration of the Cornbrook viaduct"
In: *Proc 2nd int conf on Bridge management,* pp 418–427, Elsevier

MCLAUGHLIN, D (1980)
"The repair of a cast-iron bridge over the Kennet and Avon Canal, Sydney Gardens,
Bath"
*Trans,* ASCHB, vol 5, pp 30–33

NEAVE, M A and TURNBULL, J D (1993)
"The strengthening of Bures Bridge"
In: *Proc 2nd int conf on Bridge management,* pp 397–406, Elsevier

NICHOLSON, T A (1996)
"Menai Bridge: testing, assessment and strengthening of critical members"
*Structures and Buildings,* Institution of Civil Engineers, London, vol 116, May, pp 154–162

PRESTON, W R (1989)
"Repair and maintenance of listed cast and wrought iron bridge structures"
*Conservation of engineering structures,* pp 19–32, Thomas Telford

TUBMAN, J L (1995)
"Assessment of Adelaide Bridge, Royal Leamington Spa" (Early steel latticed arch road
bridge of 1891)
*Structures and Buildings,* Institution of Civil Engineers, London, vol 110, February, pp
11–18 and discussion (1997) vol 122, August, pp 370–372

WADSWORTH, H J and WATERHOUSE, A (1967)
"Modern techniques and problems in the restoration of Marlow suspension bridge"
*Civil Engineering,* Institution of Civil Engineers, London, vol 37, June, pp 297–316